for all mankind

Foreword by Wernher von Braun

A Sunrise Book   E. P. Dutton & Co., Inc.   New York   1974

for all
mankind

*America's Space Programs
of the 1970s and Beyond*

L. B. Taylor, Jr.

*This book is for
my teenage daughter, Cindy,
for it is her generation
that must understand . . .
and make the decisions
to carry on.*

OTHER BOOKS BY L. B. TAYLOR, JR.

*Liftoff! The Story of America's Spaceport
That Others May Live: The Aerospace Rescue and
Recovery Service*
WITH KIP WAGNER
*Pieces of Eight: Recovering the Riches of a Lost Spanish
Treasure Fleet*

*Illustrations in this book are used through the courtesy of the National
Aeronautics and Space Administration, except when otherwise specified.*

First Edition

10   9   8   7   6   5   4   3   2   1

*Published simultaneously in Canada by Clarke, Irwin & Company
Limited, Toronto and Vancouver
ISBN: 0-87690-115-1
Library of Congress Catalog Card Number: 74–9109
Dutton-Sunrise, Inc., a subsidiary of E. P. Dutton & Co., Inc.
Designed by The Etheredges*

contents

v

O

list of illustrations

O

acknowledgments

I am especially grateful to Margaret Ware, a beautiful person, and to Les Gaver of NASA's Audio-Visual section in Washington for their support and cooperation in the acquisition of the photographs that accompany the text of this book. I also thank Dick Barton of Rockwell International's Space Division for providing the book jacket artwork. I appreciate, too, the efforts of John Donnelly, Assistant NASA Administrator for Public Affairs, and members of his staff for informally reviewing the manuscript prior to publication and to my good friend Chuck Hollinshead, the "Voice of NASA" at the Kennedy Space Center in Florida, for his sage advice.

I never would have made my deadline had it not been for the unselfish relinquishment of her after-hours time by Suzanne

Smith, who typed and proofed a major portion of the manuscript. I also am most appreciative of the understanding attitude of a fair-minded former boss—Walt Cloke, of Rockwell International.

I too often took for granted the many sacrifices made during the writing of this book by my wife, Norma.

But most of all, I would like to acknowledge the supreme efforts of America's dedicated men and women involved in the national space program, who, despite continuous criticism and costly budget cutbacks, have never lost the faith.

foreword

In the late 1960s, the first men landed on the moon. A few years later America established the first manned station in orbit. Now that the time has come to pay off the dividends to the American taxpayer who supported these and other space ventures, we discover the greatest beneficiary really to be our less fortunate brothers in the developing countries. We find that the true, long-term payoff of our space program is For All Mankind.

L. B. Taylor's latest book describes the cultural and economic benefits future generations will reap from communications satellites, for both international and domestic use. It talks about powerful satellites beaming audiovisual television programs into remote hamlets to help break the stranglehold of illiteracy, which, more than any other factor, still retards the progress and well-being of a substantial part of mankind. This splendid book takes us into a world where arrays of sophisticated sensors aboard orbiting spacecraft continuously monitor the status of the world's food crops, the health of the world's forests, and the natural and man-made pollution of the world's rivers. And it shows how satellites will help in the search for badly needed fossil fuels and minerals.

We also find here many new facts about the technological cornerstone of NASA's determined effort to reduce the cost of space flight, the reusable space shuttle, and how it will dramatically change the mode of future space operations. While astronauts will fly the shuttle into space and back, scientist-passengers will go along on the ride and perform their professional tasks in orbit. The shuttle is portrayed as what it really is: not just a new form of manned space flight, but a fundamental national resolve to effectively utilize the full potential of all types of future space activities. Taylor shows convincingly the seediness of the often-heard argument that we should give up manned space flight and concentrate our limited resources on the cheaper unmanned mode, which holds out enough promises to keep us busy for many years. Permitting extensive manned activities in space, while providing the most economical transportation to orbit of unmanned satellites, the shuttle emerges as the best of both worlds.

I find it gratifying that Taylor's book, for all its enthusiasm for the practical uses of space, never implies that only bread-and-butter space applications should be considered a benefit for all mankind. When he describes our plans for more extensive exploration of the planets, future research stations on the moon, or scientific schemes to glean new knowledge about the universe, his deep respect for the value of basic research unmistakably shines through. And, indeed, orbital research on astrophysical phenomena taking place in the sun may someday provide us with the key to an abundant and pollution-free energy supply right here on earth.

Once again we are entering a period when scientists and technologists are called upon to pull the rabbit out of the hat. This time they are expected to provide instant answers to our pressing problems of energy and raw material shortages in the face of overwhelming environmental problems. Basic research must precede the action of working miracles. It is the act of putting the rabbit in the hat.

Wernher von Braun

Germantown, Maryland
May, 1974

Part One

the
metamorphosis

○
Chapter 1

exploration to exploitation

The question we're asking now is: "Where do we go from here in space?" The answer is that we must put it to use to help the people more directly.

—ASTRONAUT ALAN B. SHEPARD

It somehow seems difficult to believe that the Space Age is still in its infancy. Today, satellites sail above earth routinely—relaying television programs from points all over the world, flashing timely weather pictures of developing storms, and surveying the planet's precious supply of natural resources. Robot spacecraft journey far into the solar system, transmitting unprecedented photographs of the once-mysterious faces of Venus, Mars, and Jupiter. And man ventures almost matter-of-factly into space, in orbit a few hundred miles above earth, and to the moon.

Yet historians will record the fact that the opening of this new era of infinite promise and potential occurred on October 4, 1957, when the Soviet Union successfully launched a 184-pound package of scientific instruments called Sputnik I. The United

This magnificent view of earth—from the Mediterranean Sea area to the Antarctic polar ice cap—was photographed from space by the Apollo 17 team during the final manned lunar landing mission. Almost the entire coastline of the continent of Africa is clearly delineated, with the Arabian Peninsula at the northeastern edge of Africa.

States's space endeavors were inaugurated with the placement in orbit of a tiny 30-pound satellite, Explorer I, on January 31, 1958. On October 1 of that year the National Aeronautics and Space Administration (NASA) was created, with a charter that said, in part, "that activities in space should be devoted to peaceful purposes for the benefit of all mankind."

In the few years since these beginnings the accomplishments have come at a sustained rapid pace of technological progress unmatched in history. In NASA's first fifteen years more than three hundred unmanned satellites have been sent into space, and man's expeditions were climaxed on July 20, 1969, when

Apollo 11 astronaut Neil Armstrong stepped down from his *Eagle* spacecraft and set foot upon the moon. "That's one small step for a man, one giant leap for mankind," he said, as hundreds of millions of awed earthlings listened and watched.

But while the achievements of America's national space program through the 1960s were magnificent, they were also largely exploratory. As the 1970s approached, the same public that had been thrilled by the early feats of astronauts on the moon now demanded more. We had been to the moon. Wonderful. What next? What benefits could the space program bring home to an earth plagued with pressing problems?

"We are turning from a period of space exploration to one of space exploitation," says NASA administrator Dr. James C. Fletcher. "We have come up with a balanced space program for the seventies. We are increasing our capability to do useful work in all the major areas of space activity.

"We have entered a period of increasing earthly benefits from the space program. We will, of course, continue to look outward at the universe, but we are devoting increased attention to the study of our own home planet from the vantage point of space. We can now turn our hard-won new abilities to increasingly practical use. The space program is giving us the tools to help solve many of earth's most pressing problems."

"We have moved from the era of learning how to live and work in space to a new plateau, where this nation can utilize space and its unique capabilities for expanding its horizons in science and in applications, in defense, commercial activities and in international cooperation at reduced costs," adds Dale D. Myers, former associate NASA administrator for Manned Space Flight. "The challenge facing us now is to consolidate, to redefine and to apply what we learned as we move into the era of space utilization for man's benefit."

A prime example of the exploration-to-exploitation transition was demonstrated throughout 1973 and into 1974 in the Skylab series of flights. Three three-man teams of astronauts spent from one up to nearly three months in orbit, conducting a great number of experiments designed to advance man's capabilities to live and work in the space environment, and to provide more immediate dividends for use on earth.

For instance, the astronauts trained banks of highly sensitive instruments on strips of land and sea 270 miles below to document surface features. The resulting data are being used, among other things, for mapping and studying geographic areas and geologic structures; for determining weather characteristics and patterns; for sighting crop and forestry cover, determining the health of vegetation, the types and conditions of soil, water storage, sea surface temperatures, and wind and sea conditions.

Similar information is being corrected by America's first Earth Resources Technology Satellite, ERTS-1, launched July 23, 1972. It has taken hundreds of thousands of photographs of earth conditions that are proving invaluable to hundreds of scientists in the United States and abroad as aids in the management of the earth's resources and in observing its environments.

Satellites are revolutionizing world communications, making possible the instantaneous relay of significant events across oceans and continents, and drastically reducing the cost of long-distance communications.

"Highly sophisticated weather satellites collect a broad spectrum of meteorological data on which accurate long-range—fourteen-day—weather forecasts will be based," says Dr. Fletcher. "Satellite-assisted navigation will make the airways and the sea lanes safer."

Through the use of unmanned craft, the exploration of space is providing man with a better understanding of his own planet and an opportunity to see other planets, stars, and galaxies unhindered by the earth's obscuring atmosphere. Orbiting satellites are discovering and mapping in detail the highly complex magnetosphere surrounding earth and the effect of solar radiation on earth's ionosphere and atmosphere. Other spacecraft are looking far into space to study ultraviolet, infrared, X- and gamma radiation to learn more about stars and galaxies and little understood pulsars, quasars, and black holes.

From instruments aboard Skylab and other spacecraft astronomers are gaining new insights into the workings of the sun which are leading to a better understanding of the solar system's energy balance. This could decrease the time man needs to harness more effectively the sun's enormous energies for his own uses—on the earth.

Because space knows no national boundaries, the United States is engaging in cooperative programs with countries all over the world—in fact, NASA has entered into more than 250 agreements for international space projects. One of the most exciting of these in the 1970s, and one that offers great promise of furthering world peace, is a joint venture with the Soviet Union. In 1975 separately launched American and Russian spacecraft will rendezvous and dock in earth orbit and conduct a number of scientific experiments.

Beyond this, NASA is developing a new economical transportation system to and from earth orbit, the "space shuttle." Key components will be reusable: rather than being discarded after one flight, like present hardware, they will be flown again and again, perhaps hundreds of times.

More immediately relatable to the American taxpayer are the thousands of spinoff benefits being realized from continuing space programs. Examples can be found everywhere—in the home and marketplace, in business and industry, in cities and on farms. They range from freeze-dried foods first developed for manned space flights, to advanced medical sensors that are being put to scores of practical applications in hospitals and health care centers all across the United States. New products, processes, techniques, and systems, derived initially from space programs, are finding additional uses on earth almost daily.

"During the seventies and eighties," Dr. Fletcher says, "the space effort will be serving a prime economic role as a forcing function in the development of new technology. Our space work will be a major factor in the mainstream of United States technological leadership with all that this means to the economy and our balance of trade. New technology results in new products and new industries which, in their turn, open new markets. The impact of developments of this order on employment hardly needs emphasis."

The foundation for space exploitation for the betterment of life on earth has been laid. The exploratory phase of the program is ending, and emphasis has shifted to concentrate on reaping the enormous harvest of opportunities that lie beyond earth. It is a truly historic time in the progress of man. Dr. Thomas O. Paine, former NASA administrator, summed it up well when he said:

"The opening of vast new areas for human exploration is a rare event; it occurred last in the fifteenth century on the Atlantic coast of Europe with the bold oceanic voyages of Portuguese, Spanish, Dutch, French and English seamen. Their adventurous trips beyond their traditional horizons eliminated 'terra incognita' from the globe, and brought to Europe a broad world view which has vitalized western civilization ever since. . . . I believe we are in the initial stages of a new movement which will carry broader three-dimensional perspective embracing all of the space within the earth-moon system and beyond.

"I use the analogy of the European settlement of new continents advisedly. The orbital space a few hundred miles above the continents of earth is in a real sense another continent, vast in extent, free of atmosphere, gravity and national boundaries, overlooking all of earth below. Permanent structures will be built there in which work of great social, scientific and economic value will be carried out."

America's space programs of the seventies and beyond have been designed to lead to the fulfillment of Dr. Paine's prophecy—while returning more immediate benefits to the people of earth along the way.

"We now look ahead to several decades of a highly rational use of space," says Dr. Fletcher. "The focus will be on domestic needs, and the turning of our rapidly developing space capabilities to useful work.

"We have made our new program relevant to the needs of modern America."

○
Chapter 2
the 5:15 to orbit

The space shuttle is much more than just a new vehicle. It is a whole new approach to space.

—DR. JAMES C. FLETCHER
NASA ADMINISTRATOR

Long before man ever set foot upon the moon, space experts could foresee that once the national goals of the lunar landing mission were fulfilled, a radically new, more economical, and practical transportation means to and from space would have to be found. Otherwise, with expensive launch, spacecraft, and satellite-placement costs—matched against other, higher priorities—only limited use of space could ever be made.

As the renowned writer, space prophet, and astrophysicist Arthur C. Clarke has said, "It is impossible to tolerate indefinitely a situation in which a gigantic, complex vehicle like a Saturn V [rocket] is used for a single mission, and destroys itself during flight. The Cunard Line would not stay in business for long if the *Queen Elizabeth* carried three passengers and sank after her maiden voyage. Yet Project Apollo is even more extravagant than this, because each mission also jettisons two enormously expensive pieces of equipment—the lunar module and the service module." The Apollo command ship itself does return to earth, but it is impractical to refurbish it for additional flights. A more economical and efficient system for carrying satellites, engineers, scientists and astronauts to and from earth orbit is obviously needed.

It has been generally agreed that the only practical way to lower the costs of operating in space substantially would be to develop reusable hardware. In the 1960s NASA informally began in-house feasibility studies on a reusable space vehicle system.

Preliminary data developed were then evaluated by a special Space Task Group, appointed by President Richard M. Nixon, to help define national space program objectives in the post-Apollo, post-lunar landing period. In September 1969 this special group submitted its report to the President. It recommended that the United States should develop a new transportation system to "provide a major improvement over the present way of doing business in terms of cost and operational capability; and carry passengers, supplies, rocket fuel, other spacecraft, equipment, or additional rocket stages to and from orbit on a routine aircraft-like basis."

Based on the Space Task Group's recommendation, and on follow-up studies by NASA, President Nixon, on January 5, 1972, outlined a new course for the national space program. Though not as immediately dramatic as the lunar landing goal proclaimed by President John F. Kennedy more than ten years earlier, it nevertheless would have a profound effect on space operations for years to come, perhaps for the rest of this century and beyond.

"I have decided today," Mr. Nixon said, "that the United States should proceed at once with the development of an entirely new type of space transportation system designed to help transform the space frontier of the 1970s into familiar territory, easily accessible for human endeavor in the 1980s and '90s.

"This system will center on a space vehicle that can shuttle repeatedly from earth to orbit and back. It will revolutionize transportation into near space, by routinizing it. It will take the astronomical costs out of astronautics. In short, it will go a long way toward delivering the rich benefits of practical space utilization and the valuable spinoffs from space efforts into the daily lives of Americans and all people.

"In the scientific arena, the past decade of experience has taught us that spacecraft are an irreplaceable tool for learning about our near-earth space environment, the moon, and the planets, besides being an important aid to our studies of the sun and stars. In utilizing space to meet needs on earth, we have seen the tremendous potential of satellites for intercontinental communications and worldwide weather forecasting. We are gaining the capability to use satellites as tools in global monitoring and management of natural resources, in agricultural applications, and in pollution control. We can foresee their use in guiding airliners across the oceans and in bringing televised education to wide areas of the world.

"However," the President continued, "all these possibilities, and countless others with direct and dramatic bearing on human betterment, can never be more than fractionally realized so long as every single trip from earth to orbit remains a matter of special effort and staggering expense. This is why commitment to the space shuttle program is the right next step for America to take, in moving out from our present beachhead in the sky to achieve a real working presence in space—because the space shuttle will give us routine access to space by sharply reducing costs in dollars and preparation time. . . . The resulting economies may bring operating costs down as low as one-tenth of those for present launch vehicles."

The new mode of flight and reentry, the President continued, "will make the ride safer and less demanding for the passengers, so that men and women with work to do in space can commute aloft, without having to spend years in training for the skills and rigors of old-style space flight. As scientists and technicians are actually able to accompany their instruments into space, limiting boundaries between our manned and unmanned space programs will disappear. . . . Repair or servicing of satellites in space will become possible, as will delivery of valuable payloads from orbit back to earth.

"The general reliability and versatility which the shuttle system offers seems likely to establish it quickly as the workhorse of our whole space effort, taking the place of all present launch vehicles except the very smallest and very largest.

"This new program will give more people more access to the liberating perspectives of space," the President said, "even as it extends our ability to cope with physical challenges of earth and broadens our opportunities for international cooperation in low-cost multipurpose space missions.

"'We must sail sometimes with the wind and sometimes against it,' said Oliver Wendell Holmes, 'but we must sail, and not drift, nor lie at anchor.' So with man's epic voyage into space—a voyage the United States of America has led and still shall lead."

"This decision . . . is a historic step in the nation's space program," said Dr. Fletcher. "It will change the nature of what man can do in space. By the end of this decade the nation will have the means of getting men and equipment to and from space routinely, on a moment's notice if necessary, and at a small fraction of today's cost."

Dr. Fletcher cites four main reasons why the space shuttle is important and the right next step in manned space flight in America's space program:

> First, the shuttle is the only meaningful new manned space program which can be accomplished on a modest budget. . . . Man has learned to fly in space, and man will continue to fly in space. Given this fact, the United States cannot forego its responsibility—to itself and to the free world—to have a part in manned space flight. And the space shuttle is clearly the meaningful and useful new manned space program for the coming decade.
> Second, the space shuttle is needed to make space operations less complex and less costly. . . .
> The shuttle . . . is launched vertically, flies into orbit under its own power, stays there as long as it is needed, then glides back into the atmosphere and lands on a runway, ready for its next use. With the shuttle, space operations will indeed become routine.
> Third, the space shuttle is needed to do useful things. Why are routine operations in space so important? There is no single answer to this question as there are many areas—in science, in

civilian applications, and in military applications—where we can see now that the shuttle is needed; and there will be many more by the time routine shuttle services are actually available.

Take, for example, civilian space applications. . . . The space shuttle will make it possible, in the future, to routinely launch communications and weather satellites with vastly improved capabilities—to bring education via television to remote areas, and to improve our ability to predict the weather. Also, we will have satellites that will allow us to monitor, and help us husband, our natural resources—water, minerals, and our agriculture. And perhaps with routine space operations, one could develop a global environment monitoring system, international in scope, to help control our environment here on earth.

Fourth, the shuttle will encourage far greater international participation in space flight. With the shuttle's low cost to orbit and inherent flexibility, the rest of the world—the free world at least—can work with us to launch many of their space experiments, and share with us some of the expense of space exploration. . . . Perhaps ultimately men of all nations will work together in space—in joint experiments, joint environmental monitoring, or perhaps even other joint enterprises—and through these activities help humanity unite in peace on its planet earth. . . .

We can and must look upon the space shuttle as a major investment in America's future; as the key to American power and productivity in space for the rest of the century. . . .

We need the ability to use space routinely and cheaply and extensively for scientific research, practical benefits, and national security. And for all this there is no rival, no substitute for the shuttle.

It is the logical next step forward.

The space shuttle will be totally unlike any previous rocket-spacecraft combinations in design and appearance as well as in concept. It will consist of a manned reusable orbiter, mounted piggyback at launch on a large expendable liquid propellant tank, and two recoverable and reusable solid propellant rocket boosters. The orbiter will look like a delta-wing airplane, about the size of a DC-9 jet liner. Its wingspan will be about 80 feet and its length about 124 feet. The orbiter's cargo compartment, or payload bay, will be approximately 15 feet in diameter and 60 feet long, and will be able to carry payloads up to 65,000 pounds

Blastoff: With main engines and solid rocket "strapon" motors roaring, the space shuttle will take off like a rocket and return for earth landing, like a jet transport. (Courtesy Space Division, Rockwell International)

into an orbit 115 miles above earth. This bay is about the same size as a 707's passenger section. A hatch on top of the spacecraft will swing open to deploy the payloads from the cargo area.

On the launch pad the shuttle system will stand about half as tall (206 feet) as the Apollo-Saturn rocket that launched astronauts to the moon. On the shuttle cluster the twin solid propellant rockets will be mounted on the tank at two points about 180 degrees apart. The tank, 82 feet long and 26 feet in diameter, will be mounted on the orbiter.

Total liftoff thrust from all launch engines will be nearly 8.5 million pounds. At launch the twin rocket motors, each about 13 feet in diameter and 150 feet long, will be ignited simultaneously

In roomy, comfortable "shirtsleeve" environment of space shuttle cockpit, as depicted in an artist's concept, astronaut crew pilots orbiter vehicle in space and during the return to earth. (Courtesy Space Division, Rockwell International)

with the liquid-fueled orbiter engines and will burn in parallel with the orbiter engines to an altitude of about 25 miles. The solid propellant rockets will then be detached, to descend by parachutes and be recovered, refurbished, and reused. They are expected to be capable of at least 10 reuses.

The orbiter and its propellant tank will continue into low earth orbit, possibly varying from 100 to 260 nautical miles above earth. Once the desired orbit is attained, the expendable propellant tank will be jettisoned into the ocean by the retrorockets.

The orbiter with its crew and payload will remain in orbit to carry out its mission, normally about seven days, but as long as thirty days if required. The planned normal shuttle crew of four will include a pilot, copilot, systems monitor, and a specialist who will check out the satellite payloads and deploy them in space. However, the orbiter, which can accommodate up to ten persons including the crew, is designed to permit scientists, doctors, laboratory technicians, and other men and women to work in space without special spaceflight training or pressurized space suits.

Unlike previous manned spacecraft, the shuttle orbiter will have reusable external insulation, which is likely to be relatively lightweight compacted nonmetallic tiles made from silica fibers or alumina-silica fibers. This type of insulation produces a rough, porous surface that will be coated with waterproofing to repel rain within the atmosphere.

Once the orbiter's missions are completed, it will be piloted into the earth's upper atmosphere, maneuvering as a spacecraft until it reaches approximately a 400,000-foot altitude. Here its aerodynamic surfaces and controls will take over and the vehicle will become, in effect, an airplane that can "glide" as far as 1100 miles to its base for an aircraft-type landing on an airfield up to 15,000 feet long. The vehicle can bring back from space payloads of up to 40,000 pounds. Such a reentry and landing are completely different from earlier manned spacecraft programs—from Mercury in the early 1960s through the Apollo moon landing flights—for all these spacecraft splashed down on ocean surfaces and required the enormously expensive deployment of fleets of recovery ships, aircraft, and helicopters.

Two landing sites, a continent apart, have been selected for the shuttle orbiter, one at the Kennedy Space Center on the mideast coast of Florida, the other at Vandenberg Air Force Base, about 150 miles north of Los Angeles. The Florida site will be used for research-and-development launches expected to begin in 1978 and for all operational flights launched into easterly orbits.

Facilities for all shuttle users at the Kennedy Center will be provided by NASA, largely through modifications of existing facilities built for the Apollo and other programs. Toward the end of this decade a second operational site is planned to be phased in at Vandenberg. Basic shuttle facilities required there are scheduled to be provided by the Department of Defense.

On the ground, following an orbital flight, the orbiter will undergo ground maintenance and operations, much like those carried out on conventional aircraft, and designed to have the vehicle ready for another mission within a two-week turnaround time. Such operations will include maintenance, refurbishment, and checkout; reassembly with a new main tank and boosters; payload loading; reinstallation on the launch pad; propellant loading and final systems check.

The orbiter is being designed to fly numerous earth-orbital missions. The more it is used, the more economical and practical space flight will become. When the United States launched its first satellite—that tiny Explorer I, on January 31, 1958—it cost about $100,000 per pound of payload that reached orbit. The present estimates of transporting a pound of payload into earth orbit range from $900 up. It is the goal of shuttle planners to reduce this to about $160.

As now envisioned, the shuttle will carry into space virtually

Coming home: Shuttle orbiter, following mission in space, prepares for landing on a conventional runway. (Courtesy Space Division, Rockwell International)

all the nation's civilian and military payloads in the 1980s and beyond, manned or unmanned. In so doing, NASA studies show, using the shuttle will save about $1 billion per year compared with the cost of using present-day launch vehicles. Between thirty and fifty shuttle missions are anticipated each year throughout the 1980s.

"This represents about the same mission load we are supporting today," says Dr. Fletcher. "Just doing this, the shuttle will pay for itself."

Overall, NASA has produced a "mission model" for the 1979–1990 time period, showing a total of 580 missions for the shuttle. This model has been arrived at largely through studies based on *current* space missions by NASA, the Department of Defense, and such other users as the Communications Satellite Corporation. This assumes that the pace of space flights will be maintained or increased somewhat through 1990. Actually, the model would average out to about 48 flights a year, less than one a week. NASA estimates that 262 launches would be for the space agency, 218 for the Defense Department, and about 100 for all others. Since the presidential January 1972 announcement endorsing the shuttle, NASA has received numerous responses and inquiries from possible shuttle users.

In addition to the types of missions now flown that can be achieved more economically in the future by shuttle, many new uses will be developed over the next few years.

An example, already on the drawing board, is NASA's Large Space Telescope, which may be up to 50 feet long and weigh as much as 25,000 pounds. Its scope, reportedly, will be able to look at galaxies 100 times fainter than those seen by the most powerful earth-based optical instruments. Within the solar system it will be able to provide long-term monitoring of atmospheric phenomena on Venus, Mars, Jupiter, and Saturn. Not only would the shuttle have more than the capacity needed to launch such a giant observatory in space, but it would also be able to retrieve it for repairs and for updating its instruments.

As another example, Dr. Wernher von Braun, one of the twentieth century's giants of space development, expects shuttles to help prepare the way for manned exploration of Mars. The flight, he says, probably will begin in the cargo bays of a number of space shuttles in which a Mars spaceship, its nuclear engines, its crew and supplies, will be hauled into earth orbit. There, the ship will be assembled piece by piece, the nuclear rockets ignited, and the journey begun. "I think the only way to fly to Mars is with nuclear power," Dr. von Braun says, "and because we would never want to ignite a nuclear rocket on earth, we'll have to begin the journey from orbit."

Dale Myers adds that the shuttle should usher in an era of more extensive commercial exploitation of space. Potential applications in this area, he says, including manufacturing pure serums and precision products that could benefit from the vacuum conditions of orbit. In fact, Congressman Olin ("Tiger") Teague of Texas, chairman of the House Committee on Science and Astronautics, and a strong proponent of the nation's space program since its inception, has said, "Some experts predict a $50 billion market by the end of the century for spacemade materials and biologicals."

Space visionary Dr. Krafft Ehricke, who began a distinguished career in rocketry more than thirty-five years ago along with Wernher von Braun at Peenemünde, Germany, and is now an executive with Rockwell International's Space Division in California, adds another dimension to potential uses of a new earth-to-orbit transportation system.

"With space capabilities that begin with the shuttle," he says, "we can start removing things from earth that really are alien to our environment and don't belong there. Many of these things would be better served in the greater world of space. For example, the vacuum of space is a much more benign environment for machines than is the earth's surface with its humidity, fog and corrosive salt air. . . . The industrialization of earth cannot continue to the extent needed to raise mankind's living standard significantly—that is, to the extent that would better assure peaceful conditions and the perpetuation of a plurality of cultures in a human civilization. Space—orbits now, and later, translunar regions—are favorable sites for many industrial applications."

These examples and concepts typify the great range of applications that can be carried out by the shuttle. It also is more than reasonable to assume that countless other uses will be inspired by the convenience, versatility, and economic advantages provided by this new system for getting to and from space.

Studies projecting future missions based on *current* launch schedules indicate that approximately 26 percent of the future flights are likely to be manned or man-tended. Man-tended payloads are concentrated in physics, astronomy, and processing in space because such projects involve either complicated laboratory equipment and human judgment or, in the case of astronomy, selective observations.

Of the remaining 74 percent involving delivery of unmanned missions to orbit by the shuttle, 27 percent are applications-oriented, involving earth resources, communications, and navigation; 25 percent are science-slanted, involving physics, astronomy, and planetary objectives; and 22 percent will be for the Department of Defense.

NASA estimates the total cost of the program, including all 580 shuttle missions projected to 1990, at $42.8 billion, or something over $3.5 billion per year.

Dr. Fletcher explains how some of the cost savings will be effected: "The cost of each shuttle flight will be about $10.5 million. This will reduce the cost of putting a pound of payload into space from $900 up at present to about $160, with a maximum load. And, by spreading these costs among many users, the economies are obvious.

"But this is just one cost-effective yardstick. There are many others. The shuttle will pay for itself, for example, simply as a replacement for the launch vehicles we are using now for most of our unmanned payloads. One multipurpose vehicle now will be able to perform the missions that previously required a stable of rockets.

"Critics who are opposed to the shuttle but would gladly support an unmanned program costing $2 to $3 billion a year," says Dr. Fletcher, "completely miss the point. It is unmanned programs of this sort that the shuttle is designed to serve, saving money not only in launch costs but in the way the spacecraft are built, deployed, serviced, and returned to earth for refurbishment and reuse. The shuttle is *not* primarily a manned space flight project. In fact, about three-fourths of its payloads for the foreseeable future probably will be unmanned spacecraft."

Speaking specifically on the issue of manned versus unmanned space flight, Myers says the shuttle will render the whole matter academic. "The shuttle will merge the capabilities of manned and automated space vehicles in one vehicle where now they usually complement one another in separate vehicles," he says. "One need not be blessed with extraordinary vision to see this. The issue of manned versus unmanned was dubious from the start. What is really the question is whether we want to develop the *full* potentials of space flight or only a *part* of them. Because if you eliminate man from the loop, you significantly reduce our options to explore and use space. There is no adequate substitute for man as a general sensor, calculator, evaluator, and manipulator, now or in the foreseeable future. We have proved that through every manned flight program, from Mercury to Gemini to Apollo and now Skylab. He is essential to research, development, initial operations, troubleshooting, or a combination of these. . . .

"In the 1980s, when the shuttle is operating, the scientist or engineer will have several choices. He can carry an early experimental sensor into space, adjust it, calibrate it, or tune it as an experimental project. He can use a laboratory to carry several experiments to space, operate them and return them to earth to modify them for the next flight. Or, for most flights, he will know his environment, his sensors and his mission, and he will use the shuttle as a booster stage, delivering an unmanned satellite to long-duration orbit.

"But for all of the missions, man will be able to sense, correct, repair, and recover the payloads in cases where problems develop.

"So, to sum up," Myers says, "if this nation is to make the most of the opportunities it has in space flight, then we must use *both* kinds of capabilities open to us. A paperhanger doesn't go about his task by first tying one arm behind his back. Neither, I trust, will NASA be forced into a similar measure by the idea you can explore and utilize space cheaper and better by dehumanizing it—by *not* utilizing the most reliable, easily programmed computer, the most highly sophisticated multiple degree-of-freedom manipulator, and the most versatile set of sensors ever developed—Man."

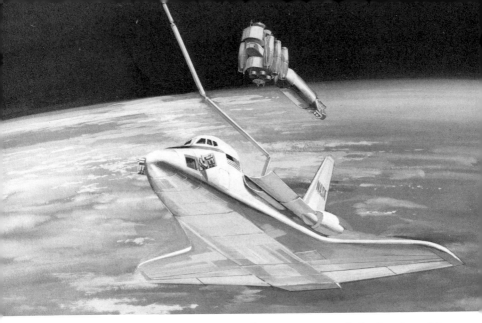

Payload recovery: With manipulator arm extended, shuttle orbiter prepares to recover orbiting satellite in this artist's concept. The key feature of the shuttle system will be its ability to retrieve payloads in orbit for repair or maintenance in space, or for return to earth. (Courtesy Space Division, Rockwell International)

"Also important," continues Dr. Fletcher, regarding the shuttle, "we can substantially reduce the cost of designing, building and operating all kinds of satellites." Actually, in launching unmanned automated payloads, the space shuttle flight crew will be able to check out the satellite in orbit to help guarantee successful operation before leaving the satellite unattended. And, if necessary, the satellite can be retrieved and returned to earth for more extensive repair to avoid a complete loss. Also, the shuttle will provide a new capability to repair a malfunctioning satellite in space or retrieve it from orbit. This alone has the potential of saving hundreds of millions of dollars.

A study of 131 satellite failures revealed that 78 were related to launch problems that could have been avoided with the more reliable shuttle vehicle. Of the remaining 53 failures, the spacecraft were inoperable or erratic and could have been returned to earth for further work if the shuttle had been available.

A direct example of this can be drawn from NASA's Orbiting Astronomical Observatory (OAO) program. A battery-charger problem that caused the OAO-1 mission to be a failure could have been corrected simply by returning the spacecraft to earth. And the shroud jettison trouble that prevented OAO-3 from attaining orbit could have been avoided if it had been launched from a shuttle already in space. Tens of millions of dollars would have been saved on these two flights alone, and no material value can be placed on the scientific data that were lost.

Precisely how the shuttle would save money on individual programs, compared with present-day costs and methods of operation, has been explained by Myers in testimony to the Senate Committee on Aeronautical and Space Sciences. Specifically, Myers was asked how the shuttle would handle a mission profile of the Earth Resources Technology Satellite (ERTS) type. This involves an earth observation spacecraft placed in low polar orbit for observation of meterology, oceanography, pollution, and natural resources phenomena.

"For Thor Delta [rocket] carriers," Myers said, "launching one satellite per year to an altitude of 450 nautical miles between 1979 and 1990 would require twelve [launch] vehicles and twelve spacecraft. The average cost per satellite in orbit would reach $75 million.

"Since the shuttle provides a benign environment, the satellite would be redesigned for retrieval, refurbishment and low cost. This would result in doubling the weight from 2590 pounds for Thor Delta, to more than 5000 pounds for the shuttle. But the capability of the shuttle to retrieve the satellite from . . . orbit permits a reduction in the number of spacecraft since some would be returned to earth, refurbished and redeployed.

"Instead of buying twelve new satellites, the same missions can be flown with the same results with only two new spacecraft, each of which would be refurbished five times. Thus the cost per satellite in orbit would come down to $46.8 million, or a saving of 38 percent."

Myers also compared the cost of handling communications satellites with current Titan IIIB/Centaur launch vehicles, and with the shuttle. Using the present expendable boosters, he estimated the cost would average $25.8 million per mission. Performing the same program with the shuttle reduces the average cost to $14.9 million, or a saving of 42 percent.

The beauty of this new space transportation system, however, is its inherent versatility, its multiple capabilities. So many things can be done with it in so many ways. "It can do so many combinations of things," Myers explains. For example, "it can take more than one payload up, or it can take something up, deploy it, change orbits, and pick up a satellite for return to earth. In this way it can perform more than one mission on a single flight. This is impossible under the present one-shot launch system."

Overall, assuming the 580-flight mission model for the twelve-year-period 1979–1990, NASA projects total savings of $13.4 billion.

It is important to emphasize also that the specific missions that justify the shuttle in this economic respect are those that could and would otherwise be justified on their own merits with conventional launch vehicles. In other words, even if the shuttle never left the drawing boards, these flights probably would take place anyway. The shuttle, however, makes them more effective and less expensive.

Justification of the space shuttle is not based, however, on the details of mission economics alone. The fundamental reason for developing this new transportation system is the necessity to have a means for routine quick reaction and economical access to space and return to earth to achieve the benefits of the scientific, civil, and military uses of space that will be important in the decade of the 1980s and beyond.

NASA estimates that development of the shuttle system, over a six-year period, will cost about $5.15 billion. This includes development, test, and procurement of two orbiters and two boosters through 1978. Additional investment costs for procurement of production flight hardware is estimated at about $1 billion, on the "reasonable assumption" that the initial inventory will include three production orbiters, two refurbished orbiters, and the initial production boosters. The facilities for development, test, launch, and landing capability are expected to run about $300 million, for a total shuttle development cost of about $6.45 billion over a period of several years.

"We have worked very hard on design changes to bring

these costs down to where annual funding will not exceed $1 billion," says Dr. Fletcher. "Some opponents of the shuttle misuse their figures and come to the false conclusion that the program will cost $30 to $40 billion over the next two decades. They do it by inflating development costs; by saying the number of launches per year must be greatly increased, which is not true; and by incorrectly including the cost of payloads as part of the cost of the shuttle.

"Their figures are wrong and their logic is wrong. It is against common sense to add the operational costs of the shuttle to its development costs. The cost of using the shuttle to carry out a space mission should be added to the cost of the mission. Actually, we will be saving money, not spending it, every time we use the shuttle for a space mission.

On July 26, 1972, NASA announced the selection of the Space Division of Rockwell International for "negotiation of a contract to begin development of the space shuttle system. As prime contractor, the company is responsible for design, development, manufacturing, test, and evaluation of two orbiter vehicles and support for the initial orbital flights, and for integration of all elements of the program."

The contract is for $2.6 billion over a period of about six years. More than half of this total is being subcontracted to as many as 10,000 firms in nearly every state, and employment at Rockwell's Space Division, primarily in Southern California, is expected to build to an estimated total of 9000 persons assigned to the program by 1975–1976.

Among the nation's foremost aerospace companies that have joined the shuttle development team are McDonnell-Douglas Astronautics Company, St. Louis; Grumman Aerospace Corporation, Long Island, New York; the Convair Aerospace Division of General Dynamics, San Diego; the Republic Division of Fairchild Industries, Inc., Long Island, New York; Honeywell's Aerospace Division, St. Petersburg, Florida; IBM's Federal Systems Division, Owego, New York; and American Airlines' Maintenance and Engineering Center, Tulsa. Rockwell International's Rocketdyne Division in Canoga Park, California, which has designed, developed, and manufactured the rocket engines that have boosted more than 80 percent of America's spacecraft and satellites into space, was also selected by NASA to build the shuttle orbiter's main engines. The Martin-Marietta Corporation, Denver, was selected by NASA to design, develop, test, and evaluate the external tank for the space shuttle under a $100-million-plus contract. All of these firms have long been involved in the national space program.

Thus, in addition to its primary contributions to our overall technological capabilities and increasing contributions to a continuing strong posture in space exploration and exploitation, the shuttle will have a significant impact on the national economy. At the peak level of activity, it is estimated the program will generate an estimated 50,000 jobs.

The New York Times, in February 1972, editorialized: "The space shuttle represents a major investment in the future. If implemented successfully, it will radically alter the economics of space activities and provide dividends that should continue for decades to come. More immediately, appropriation of the needed sums—averaging almost a billion dollars annually for the next six years—will revitalize a major branch of American technology which could literally disintegrate if no such project were undertaken [in] this decade."

If current schedules hold, the first experimental horizontal flight of the orbiter will be in 1977 and the first manned orbital flight about a year and a half later. It is hoped that operational status can be achieved late in 1979 or 1980. To date, the program is on schedule and proceeding well.

While the space shuttle will have the versatility, depending upon payload weights, to achieve orbits stretching from 115 perhaps up to 500 miles above earth, many missions will need to range farther out in space to perform most effectively. In fact, NASA estimates that in the period 1979–1990 about 50 percent of the satellites will require a higher orbit than those attainable with the shuttle alone. About half of these higher orbits will be geosynchronous—22,000-plus miles above earth. The other half will include flights into higher-altitude polar orbit and orbits beyond the reach of the earth's gravitational pull.

As envisioned in present planning, a special new reusable vehicle would be used to support all shuttle missions that will extend beyond low earth orbit. It is called the space tug. It will feature a highly efficient lightweight ground-based propulsion stage that will be unmanned and will be launched, with payload, from the shuttle's cargo bay. Because the tug will be operated in an unmanned mode, highly reliable docking and retrieval techniques must be developed to meet payload delivery, return or servicing requirements. Manned assistance may be remotely employed for such complex operations as docking, but essentially the tug will be capable of independent docking with satellites or other payloads in high orbits which will have compatible mechanisms.

The present tug concept is for a vehicle 35 feet long, 15 feet in diamater, and weighing approximately 60,000 pounds when fully loaded with fuel. It would be capable of carrying 3000 pounds round trip from earth to geosynchronous orbit.

Orbital workhorse: Artist's rendering of the versatile space tug, right, deployed from shuttle orbiter to place dual payloads in high-energy earth orbits. Tug also could be utilized for injecting planetary mission payloads.

A shuttle-tug system offers a variety of flight modes to meet practically any conceivable orbital mission requirement. A typical flight profile could be as follows: The tug, with its automated payload, is placed in the shuttle orbiter cargo bay at the launch site. The shuttle is launched into a parking orbit. Next, the tug and its payload are deployed and flown to geosynchronous orbit for deployment of the payload. If it is a round-trip mission, the tug then seeks out and docks with another payload that has been in orbit for some time and needs to be returned to earth. The tug returns with this payload to the shuttle orbiter in low earth orbit. Following this rendezvous and docking, the whole package—tug, payload, and orbiter—returns to earth.

Such a round-trip geosynchronous orbital flight for the tug would take from three to six days, depending on the locations of the payloads. Other missions would include the tug's deploying a payload only and returning to the orbiter without a spacecraft (payload placement); ascending to synchronous orbit or any other high orbit without a payload and returning with one (payload retrieval); or ascending with a particular device, servicing a satellite, and returning (in-orbit servicing).

NASA estimates the cost of the tug's development at between $650 and $750 million, and such funds probably will not be available until after the shuttle funding peak in 1977. The tug could not be fully operational, therefore, until about 1984 or 1985, four to five years after the shuttle is flying on a regularly scheduled basis. To fill the need for propelling payloads to higher orbits during this period, a number of interim tuglike concepts are being studied.

Whichever way the planners decide their objectives can best be achieved, it has been definitely ascertained that a standard space tug is an element essential to meeting the future payload placement, retrieval, and servicing requirements of the Air Force and NASA, as well as other civil and international users. With such a tug the space shuttle will have a fully operational capability that will extend to all orbits above the planet, from near earth to geosynchronous altitudes.

After nearly three years of discussion of possible European participation in the development and utilization of the space shuttle, the European Space Conference in December 1972 endorsed the development of a sortie or space laboratory to be flown in the shuttle. It is to be fully funded, designed, and developed under the aegis of the European Space Research Organization (ESRO). In the spacelab, which will fit snugly in the shuttle's cargo bay, investigators will have direct access to their experimental equipment. They will perform experiments in astronomy, space physics, life sciences, earth observations, material sciences and manufacturing, communications and navigation, and advanced technology.

Recent agreements between the United States and European nations on development of the spacelab for use with the shuttle call for a broad exchange of technology and support—including plans for European astronauts to fly as shuttle crew members when the laboratory is carried into space.

NASA has invited participation by European scientists in a number of study groups to develop recommendations for early shuttle use and to define the interface and support requirements these uses will impose on the spacelab and the shuttle orbiter.

"The space shuttle is a program tailored to the times," says Dr. Fletcher. "It is a worthy challenge for progress-minded people. It is in line with our traditions as a pioneering nation. When the chance came to bind America together with the transcontinental railroad, we didn't say no because we already had the Pony Express. Let's not say no to the shuttle just because we have old technology rockets that served our needs in the last decade.

"Let's say yes to the shuttle . . . because it is America's best bid—America's only bid—for a place on the space frontier in this decade.

"We can and must look upon the space shuttle as a major investment in America's future, as the key to American power and productivity in space for the rest of the twentieth century.

"Scientifically and technologically," says Dr. Fletcher, "we are laying the foundations for America's greatest epoch. Touching upon virtually every area of human endeavor, the space program offers one of our best hopes of making Spaceship Earth a safe and desirable place to live and work for future generations."

○
Chapter 3

home is 270 miles up

There was trouble right from the start. Serious trouble. Sixty-three seconds after liftoff, on May 14, 1973, as the massive Saturn V rocket thundered skyward, at an altitude of about 40,000 feet, approaching the region of greatest stress and vibration, an aluminum micrometeoroid shield on Skylab I inexplicably ripped away.

At first this went unnoticed by ground controllers, and the huge Skylab experimental space station, the first ever launched by the United States, soared on into a nearly perfect orbit about 271 miles above the planet. Precisely on schedule, signals from an onboard computer triggered release of a metal covering that had protected part of the station-laboratory during launching and flight. Solar telescopes were shifted into position and solar panels were deployed atop the telescopic unit.

30

Striking view of Skylab cluster was taken by astronauts aboard Apollo spacecraft during an inspection "fly around." Telescope mount is at left, and solar panels extend from the laboratory workshop.

All seemed to be going well and the first Skylab team of astronauts, veteran Charles ("Pete") Conrad, Paul Weitz, and Dr. Joseph Kerwin, were set to follow the lab into space a day later, rendezvous with it, and make it their home in orbit for twenty-eight days.

But less than an hour after liftoff the ominous signs of the trouble were apparent on earth. Flight controllers at NASA's Johnson Space Center south of Houston failed to receive radioed data confirming that the two largest solar panels had deployed like wings from each side of the laboratory. The mechanism that unlocks them for unfolding somehow must have malfunctioned when the shielding tore away a minute and three seconds after launch.

Without these two key panels, which convert sunlight to electricity, Skylab's electrical capacity would be cut in half. Many of the scientific experiments planned for the eight-month-long program might have to be sharply curtailed or eliminated.

As the minutes ticked by and the astronauts waited anxiously for their launch, the troubles worsened. Flight controllers next saw the disturbing data indicating that the aluminum shield was missing. It was supposed to have deployed as a sort of super heat reflector and umbrella, serving as a barrier to high-energy cosmic particles and, with its coat of white paint, it was to reflect the sun's intense rays.

Without this shield the space station's unprotected metal hull heated up to nearly 300 degrees during it's first twenty-four hours in orbit. Inside Skylab, where the astronauts were to live, the bombardment of solar heat was too great for the air-conditioning system and the interior of the station became almost ovenlike, with temperatures shooting up to an average of 120 degrees and in certain spots to as much as 190 degrees.

It was obvious the astronauts could not live and work in such a sweltering environment for even a short time, much less twenty-eight days, so their launch was postponed while engineers and officials studied the problems. Aside from the uninhabitability of the Skylab interior, it was feared also that the heat could cause some materials to emit carbon dioxide and carbon monoxide, thus poisoning the atmosphere. It was believed that some of the photographic film and medicines had probably been damaged by the heat and that some of the food sent up for the crew had probably spoiled.

To many, the entire $2.5-billion Skylab program looked doomed.

Conceived long before the first manned landing on the moon, Skylab's mission objectives were to "routinize" the phenomenon of man in space, lead to the inevitable day when man will live and work on advanced space stations for long periods, and channel a harvest of space benefits for the betterment of life on earth.

"Skylab is a new dimension in space research and technology," says William Schneider, NASA's director of the program. "It is the first major application of what has been learned so far in space."

Specifically, NASA says, Skylab was targeted to "carry out a wide range of experimental investigations and to gain a better understanding of the requirements for a permanent manmade platform in space." Emphasis was to center on three prime areas of interest: a series of medical experiments aimed at determining the effects of extended space flight on man; a package of earth survey experiments; and a group of high-resolution solar astronomy experiments to give scientists a look at the sun's activity undistorted by the constant shimmering haze of earth's atmosphere.

Aside from the rockets that would lift the men and machines to orbit, there are five essential hardware elements in the Skylab cluster: an orbital workshop, an Apollo Telescope Mount (ATM), an airlock module, a multiple docking adapter (MDA), and the Apollo spacecraft.

The workshop itself has the outside appearance of a stubby silo. Actually, it is a completely refurbished S-IVB third stage, much like those used on the Saturn V launch vehicles to help propel Apollo spacecraft to the moon. It is cylindrical, 48 feet long, 21 feet in diameter, weighs 76,000 pounds, and encompasses about 10,000 cubic feet of space.

It differs, however, from earlier S-IVBs inside. Its propulsive engine has been taken off and the complex plumbing used to contain and mix high-powered propellant mixtures stripped out.

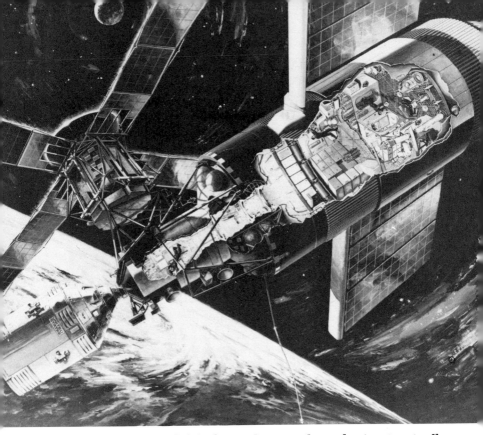

Artist's cutaway of Skylab cluster shows earth-to-orbit ferrying Apollo spacecraft, left, connected to an airlock module and multiple docking adapter, which lead into the main orbiting workshop area, right. Cartwheel-shaped object above docking adapter is the Apollo Telescope mount.

The entire tank has been "interior decorated" to include many comforts of home and tools for work.

It was in this workshop, built by McDonnell-Douglas Astronautics, that the astronauts were to spend most of their time while in orbit. The tank had grid-pattern floors and ceilings separating living and laboratory working areas into a two-story arrangement.

In the aft end solid partitions divide the crew quarters into individual sleeping compartments, a dining area, a "bathroom" and work/experiment quarters. At the other end, occupying perhaps three-quarters of the room, is the lab. Here much of the in-flight work was to take place. The multiple docking adapter, with its 17-foot length and 10-foot diameter, looks like a gigantic oil drum. It has a docking port for the Apollo spacecraft, houses the control panel for the Apollo Telescope Mount, provides a window for earth-resources viewing, and has other experimental capabilities and stowage areas.

The airlock module, built by the McDonnell-Douglas Astronautics, is the nerve center for the Skylab cluster. The connecting link, or pressurized passageway, between the workshop and MDA, it is the focal point for all extravehicular activity (EVA), or "space walks," during missions, and contains the main communications/data transmittal links, environmental/thermal system, and the electrical power control system.

The Apollo Telescope Mount looks somewhat like an automobile radiator fan—except that it is 13 feet long, weighs 22,000 pounds, and with its four solar panels (the fan blades) extended, measures 1200 square feet in total area. It houses the solar experiments as well as a solar array to provide electrical power in flight.

The fifth critical element of the Skylab cluster is the Apollo Command and Service Modules (CSM), designed and manufactured by Rockwell International's Space Division for the manned-lunar-landing program and extensively modified for its new job.

The whole arrangement—workshop, multiple docking adapter, airlock, and Apollo—fans out to a configuration 118 feet long and weighs 181,300 pounds, not counting the telescope mount. Total work area is close to 13,000 cubic feet, or about 1600 square feet, which is comparable to a small three-bedroom house. Moonbound Apollo spacecraft, by contrast, had roominess comparable to a Volkswagen van.

On paper the mission sequence was to have gone like this: the Skylab workshop, telescope mount, multiple docking adapter, and airlock module—all neatly cocooned by protective shielding—would be launched into a near-circular orbit about 270 miles above earth. The flight would be aboard a two-stage Saturn V rocket.

Assuming all went well, a smaller Saturn IB rocket, capped by a modified Apollo spacecraft containing the first team of Skylab astronauts, would be boosted next day into orbit about 100 miles up. From that vantage point, using techniques perfected during the Gemini and Apollo programs, the crew would rendezvous and dock with the Skylab cluster and spend twenty-eight days circling the earth. They would then return home and the workshop would be shut down for sixty-one days.

Late in July 1973 a second team of astronauts—veteran Alan Bean, who had explored the moon with Pete Conrad on Apollo 12, and space flight rookies Owen Garriott and Jack Lousma—would fly up and man the station for almost two months. Plans called for them to return and be replaced, after the station had been inactive for thirty-four days, by a third crew. Gerald Carr, Edward Gibson, and William Pogue would man Skylab for a final period of two months or longer.

Upon their return the orbiting hardware, its job completed, would be deactivated. The entire flight sequence, including three separate manned missions—from first launch to final splashdown—was intended to require eight months.

NASA expected Skylab to help determine whether the astronauts would be able to carry out their tasks satisfactorily in the weightless vacuum of space over relatively long periods. Isolated from civilization, would they get tired and irritable? Homesick? Would such factors affect their performance, and if so, how much? Does a long time in orbit subject the human system to serious danger?

Planning included putting the crews through carefully devised exercise routines to measure metabolic and cardiovascular systems in depth. Investigations of space effect on nutrition, hematology, immunology, neurophysiology, and pulmonary function, among others, also were to be carried out.

Scientists are primarily interested in identifying the precise mechanisms that alter human body chemistry in the absence of gravity. Not only is this knowledge important in determining how to control adverse reactions during long manned space flights, it can also contribute to an understanding of life processes which is basic to treating human illnesses.

More easily understood by the program's backers—the

American taxpayers—would be the information gathered from the Earth Resources Experiment Package (EREP). This phase of the Skylab mission has near-term application to help man better manage his environment's rich resources with new tools.

Specifically, there were to be six earth-resources experiments in which remote sensing equipment would map geographic and weather features, crop and forestry cover, health of vegetation, types of soil, water storage in snow pack, geological features associated with mineral deposits, sea surface temperature, wind and sea conditions, and the location of probable feeding areas for fish.

The Skylab flight plan was designed so the cluster would fly over the entire United States, except Alaska, at five miles per second, and it was to cover about 75 percent of the earth's surface. Its flight track was to be repeated every five days during the lifetime of the program.

The Apollo Telescope Mount may prove the greatest solar astronomical advance since Galileo's invention of the telescope in 1609. "The study of the sun," says Program Director Schneider, "goes back to the beginnings of civilization, when prehistoric astronomers first used its motion to predict the seasons and tell the best times for planting and harvesting. Their successors, modern solar astronomers, seek to understand and explain the remarkable phenomena within and around the sun itself. In part, this is scientific research in its purest form, but there is also a strong awareness that better knowledge of solar processes may lead the way to new means for generating and controlling energy for use on earth."

There are several major experiments on ATM, aimed at obtaining solar measurements in the extreme ultraviolet and X-ray portions of the electromagnetic spectrum, which cannot penetrate the earth's atmosphere, and also gathering pictures of the sun's corona in the white-light portion of the spectrum.

In astrophysics answers have traditionally been sought to fundamental questions regarding the nature of the physical universe, leading to the discovery of important concepts—the passage and measurement of time, the seasons, the size and shape of the earth, its place in the solar system and relationship with the rest of the observable universe. In addition, the study of matter in previously unknown states and of processes too exotic to occur naturally in earth's environment—the generation of thermonuclear energy in the center of stars, for example—has had a great influence on the growth of other physical sciences.

The sustained weightless environment also affords a unique laboratory for a number of corollary experiments aside from those in the medical, solar astronomy, astrophysical, and earth resources categories. One of the most intriguing series of tests was designed to study the effects of weightlessness on such manufacturing techniques as the casting of perfect spheres, the growth of pure crystal structures, and the development of foamed high-strength materials—all of which cannot now be produced on earth.

Another fascinating and highly unusual Skylab project— cosponsored by NASA and the National Association of Teachers

Studying the sun, astronaut Owen Gariott, science pilot for the second manned Skylab mission, operates the Apollo Telescope Mount console in the multiple docking adapter area of the orbiting space laboratory.

of Science—allows active participation by talented high-school students all across the United States. To stimulate youth interest in science and technology, proposals were solicited from the students for experiments and demonstrations for actual flight on the earth-orbiting missions.

Of the more than thirty-four hundred high-schoolers responding, nineteen finalists were selected to take part in the program. Some of the experiments chosen represented surprisingly complex subject matter. Among the winning entries were chicken embryology in zero gravity, photography of libration clouds, X-ray emission from Jupiter, and search for pulsars in ultraviolet wavelengths.

One student whose idea was chosen suggested the observation of volcanoes on earth from space with infrared cameras and equipment. An increase in heat as measured by such detectors might prove to signal a potential eruption. Data so obtained might make it possible to predict volcanic activity, thus saving lives and property in many areas of the world.

The more than eighty experiments in the entire Skylab mission are being conducted by some six hundred principal investigators and coinvestigators.

Because of the roominess of the workshop and experience gained from past space flight, it was believed more "conveniences of home" could be attained on Skylab. For one thing, the meals would be better, more appetizing, and much more varied than anything offered on Mercury, Gemini, or Apollo through the 1960s and the early 1970s. The astronauts would carry reconstituted freeze-dried, wet pack, and frozen foods, and many items could be electrically heated in serving trays.

In a zero-gravity environment crew members would wear comfortable cotton overalls, donning their space suits only for extravehicular activity—walks in space outside the spacecraft-workshop cluster. Temperatures in Skylab were to range from 60 to 90 degrees. The men would rest and sleep in private lightweight sleeping-baglike restraints designed to keep them from floating about. For the first time on any space flight they would be able to shower aboard ship under a pressure spray, which would force water over them (otherwise the drops would dance about randomly in the weightless atmosphere). Bars of soap were to have built-in steel disks that would cling to a magnet, and a new kind of toothpaste would be swallowable.

But in mid-May 1973—after years of planning, designing, hardware development, systems integration, test flights, preparation of scientific instruments and experiments, and arduous crew training—all seemed lost before even a single Skylab astronaut reached the orbiting laboratory.

Space engineers are a peculiar breed, however: "impossible" isn't in their dictionary, and no problem is insurmountable. Through the years, from the earliest shaky space flights to Apollo, they have pulled off a number of minor miracles. Now they set out to solve the Skylab trouble.

First, ground controllers quickly radioed instructions ordering Skylab to change its angle of exposure to the sun. Shortly thereafter, the spaceship's internal temperature stabilized at just under 100 degrees—still too hot to permit crewmen to perform inside the laboratory.

Then, working around the clock, they sought a way to shade the laboratory from the sun's blistering effects, which were driving temperatures inside the spaceship to uninhabitable levels. They speculated that if half or even less of the spacecraft area facing the sun could be covered, it might be enough to allow astronauts to board.

The launch of the Conrad-Weitz-Kerwin team was delayed several days to allow time not only to come up with possible solutions, but to test them as much as possible under the circumstances. Engineers finally settled on a plan to have the crew erect in space an aluminized awning, of Mylar, an extremely thin, mirrorlike plastic foil, 22 by 24 feet, over the Skylab roof.

For days technicians and skilled craftsmen worked frantically to ready the special awning, and astronaut teams practiced its deployment to prove, as well as they could on earth, that it was feasible to erect it in space. The Apollo spacecraft was jammed with all sorts of repair tools and the awning, and the flight was rescheduled for May 25.

Launch was perfect and within seven and a half hours the Conrad team maneuvered their spacecraft through varying orbits to close in and catch up with the Skylab cluster. They flew in tandem with it for a while, visually inspecting it. Then Weitz, wearing a full pressure suit, leaned out of the open Apollo hatch and tried to release the jammed solar power wing with special cutting tools on the end of ten-foot poles. A narrow piece of aluminum—"one lousy bolt," Conrad called it—appeared to be holding it close to Skylab's hull, and the solar wing would not budge. (The second solar panel wing apparently had been lost in flight at the same time as the micrometeoroid shield ripped off.)

After checking the interior of the workshop with a chemical detection device, to make sure no poisonous gases had been unloosed owing to the intense heat inside and its affecting equipment, the crew entered their home away from home, finding the temperature tolerable if not the most comfortable. The next day the astronauts successfully deployed their makeshift sunshade, and although it did not fully unfold it protected enough of the workshop roof from the sun to send temperatures inside plummeting almost immediately. When the news that the "umbrella" was working was relayed to earth, the Anniston, Alabama, *Star* headlined: MARY POPPINS LIVES!

Without the full electrical power provided by the solar panels, however, the two follow-on flights probably would have to be canceled. A few days later, therefore, a second major effort was made to cut the balky bolt loose from the solar panels. Conrad and Kerwin, standing outside the aft end of the workshop, assembled a twenty-five-foot aluminum pole from five sections and attached a cable-cutter to one end of it. They maneuvered the pole to clamp the cutter jaws to the debris on the solar wing beam. Tugging mightily, the astronauts snapped the cutter's blades shut, severing the stuck aluminum strip, and the jammed wing sprang out to full deployment.

The additional electrical power, lowered temperatures in the workshop provided by the sunshade, and old-fashioned "Yankee ingenuity" had saved the mission.

"I think the whole Skylab performance is a reaffirmation of the American ability to meet challenges and conquer them in the face of considerable odds," said NASA's Dale Myers.

Chairman Frank E. Moss of the Senate Space Committee, said, "With other Americans I have watched breathlessly to see what was done to restore the damage and to continue the invaluable series of experiments to be performed in Skylab. Fortunately, it appears now that nearly all of the planned experiments can be executed and what appeared as a potential failure now could be a crowning example of man's ability to work in space and meet unexpected problems with adequate solutions."

The repairs made, the first Skylab crew completed its twenty-eight-day, 11.5-million-mile mission and returned safely to earth on June 22, 1973. Overall, from what had been feared a lost mission, the astronauts were able to complete 88 percent of the Apollo Telescope Mount experiments, obtaining 30,000 pictures of regions of the sun; 88 percent of the earth-resources experiments, including photographic imagery of 182 sites in 31 states and 9 nations; 90 percent of the medical experiments; and 100 percent of all other planned experiments.

About the solar information collected, Dr. Robert Noyes of Harvard, one of the principal investigators, commented, "The data as we've looked at them frankly exceed our wildest expectations. We have seen entirely new and different phenomena." At a press conference in Houston held in June 1973 another scientist involved in the analysis of the returns from the Apollo Telescope Mount experiments said, "The total sum of these observations will far exceed what man in all recorded history has ever observed of the corona. So in that one particular experiment we're seeing an increase in our knowledge in observation of the corona that's unparalelled."

Scientists learned, for example, that the corona is much more turbulent and dynamic than they had expected—the result of an interplay between the magnetic field and charged particles emanating from the sun. This produces the many eruptions of energy known as solar flares and the spiral arms of charged particles, which resemble cyclones, that can be detected in the corona. "We have accumulated more data," said Dr. Robert MacQueen, a principal investigator of the sun's corona, "than could be obtained during eighty-four total eclipses from the earth."

Scientists were equally enthusiastic about early returns from the earth-resources experiments, which produced photography containing far more detail than pictures previously beamed to earth from unmanned satellites, including ERTS-1. "Photos of the United States are the best we've ever seen," said Conrad. "You can see railroad tracks and the ties on the railroad tracks. They're about six inches across. That's pretty good from an altitude of two hundred and seventy miles!"

"The sensor operations and the data we're going to have back from the principal investigators are going to clearly demonstrate how useful the scanner cameras and the microwave systems are for earth resources," said Dr. Verl R. Wilmarth, of NASA's EREP program office at the Johnson Space Center. About 12,800 frames of data imagery from the various camera systems were obtained. Such imagery, covering all types of ground conditions—snow, croplands, urban areas, pollution, and cloud buildups, among others, over most of the earth's populated regions from the Canadian border to the tip of Argentina—will be studied by the scientific investigators for months, perhaps years.

The Skylab crew was able to focus its equipment on targets of intense interest to scientists. Some examples:

- pictures were taken of Central Florida to study the effects of explosive urban growth caused by the opening of Walt Disney World near Orlando, Florida;
- pollution and weather patterns in the Great Lakes region were photographed for study and use by the surrounding states;
- cameras were aimed at Oregon's Columbia River Basin to analyze the depth of snow cover; such data are expected to permit more accurate forecasting of water runoff;
- states throughout the Southwest were photographed to obtain more information about the area's potential for earthquakes and to spot undiscovered mineral deposits.

"The earth resources data that we are getting back from these investigators . . . will have a direct use," says Dr. Wilmarth, "certainly by many regional planners, urban planners, agriculturalists, forest inventory people—as a means of getting out a better system for them to acquire their information to do their inventory and do their management planning that goes on with all of these activities."

The first Skylab flight was also a milestone in the medical history of the United States manned spaceflight program. The crew's performance during what was then the longest mission in space was outstanding. Near the close of the four-week flight crewman Joe Kerwin, the first physician to fly in space, said he, Conrad, and Weitz were all in excellent condition. "This gives me tremendous encouragement about future long-duration flights," Kerwin noted.

Kerwin conducted exhaustive tests on all three crewmen in his onboard medical laboratory, and said that the long exposure to weightlessness had apparently caused "some body changes in some areas and none in others." Doctors examining the astronauts after their return to earth confirmed this but said all such changes were consistent with the predicted adaptive processes anticipated by space program medical specialists.

The overall assessment was that from a medical standpoint man had proved that he can operate efficiently in space for a relatively long time without endangering his health. In fact, Conrad commented, "I'd say very definitely that the average man or woman could fly in space."

Summing up the first manned Skylab mission, Dr. Fletcher said, "I ought to read a message from the President . . . addressed to astronauts Conrad, Kerwin, and Weitz: 'The successful completion of the first mission of Skylab is a source of intense pride for the American people. You have demonstrated that, just as man can conquer the elements of earth, he can cope with the exigencies of space. You have given conclusive evidence that, even with the most advanced scientific and technological support in the world, the courage and resourcefulness of good men are still central to the success of the human adventure.' "

On July 28, 1973, about a month and a half after completion of the first manned Skylab mission, the second crew—Alan Bean, Jack Lousma, and scientist-pilot Owen Garriott—were launched from the Kennedy Space Center on a fifty-nine-day mission. In the Apollo spacecraft they carried with them to earth orbit a record cargo, including experimental equipment and replacement items for the orbiting laboratory, 1900 pounds in all.

Bean, Lousma, and Garriott set a bookful of new space records and doubled the information—on earth, sun, and man's health in space—gathered by the Conrad team.

More than a dozen nations in Europe, Africa, Asia, and South America asked the second team of Skylab astronauts to use their cameras and sensors to explore problems in soil use, water conservation, crop diseases, ocean fishing, and land-use surveys. For example, considerable data were collected on passes over Israel following that nation's request for an inventory of major soil types, crop diseases, and insect infestation.

Earth-resources data gathered over Africa provided information on the location of surface water and vegetation, which helped scientists to understand better a severe drought that was affecting the continent during the summer of 1973.

On September 25, 1973, precisely on schedule, Bean, Garriott, and Lousma returned to earth after a record-setting fifty-nine days in orbit during which they traveled more than 24 million miles. Doctors were overjoyed by their remarkable physical condition after the long journey. This immediately buoyed the hopes of scientists that the human body may adapt to weightlessness without continuous physical deterioration. This would mean that flights of unlimited duration, such as manned missions to Mars and other planets, are feasible from the human endurance standpoint.

Overall, the work on the second manned Skylab mission was prodigious. The crew returned with 77,600 pictures of the sun's corona, 14,000 pictures of the earth, and more than 18 miles of magnetic-tape data.

Using the vast array of earth-study equipment aboard Skylab, the astronauts made approximately forty observation passes over the major portions of the globe. Space cameras photographed snow fields in Switzerland, geological features in Spain, and pollution in Germany. Also covered were the movements of tropical storms, locust swarms, crop growth, and water resources. Data on geology, forestry, and agriculture were gathered in passes over the United States, Central and South America, and Asiatic areas.

In their continuing study of the sun, the crew recorded a solar flare which officials said equaled 100 million times the energy of a one-megaton atomic bomb. In all, 50 percent more solar photographs were taken than were planned. Valuable data were col-

Upside-down service: In the weightlessness of space aboard Skylab, astronaut Dr. Joe Kerwin gives a physical examination to flight commander Pete Conrad.

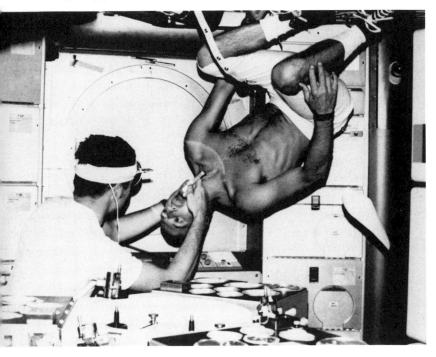

Alone in space, outside the confines of Skylab, astronaut Jack Lousma deploys a twin-pole solar shield to help shade the orbiting workshop from the sun.

lected during a time when the sun put on a brilliant display of explosive flares and other turbulent phenomena.

Bean, Lousma, and Garriott, practically on a daily basis, continued the life science experiments, investigating the effects on man as they exposed themselves to long durations in the weightless space environment.

As an experimental "bonus" Lousma fired up an electrical furnace to form new metal alloys that will not blend on earth because of gravity. This experiment originally had been scheduled to begin on the final Skylab flight, but Lousma was familiar with the process of setting it up. The work shed new light on the action of materials in a gravity-free environment and could lead the way for manufacturing on a large scale in future space laboratories.

All of the photographs, computer tape, and other data gathered during the Skylab missions undergo extensive analysis by 137 principal investigators and their technical staffs in the United States and in 18 foreign nations. Individual research projects may take up to 18 months or longer.

Preliminary reports from Skylab investigators indicated some possible discoveries of major importance. Research efforts in geology, for example, could pinpoint potential new sources of coal, oil, and essential minerals. M. L. Jensen of the University of Utah stated in September 1973 that his study of Nevada geology had uncovered a region likely to contain a significant mineral deposit. Partial information on potential geothermal energy resources in the western United States and Mexico also has been collected on Skylab.

James H. Wray of the U.S. Geological Survey is attempting to calculate population growth since the 1970 census using Skylab data. During the first two manned flights, thirteen United States urban areas, stretching from Asheville, North Carolina, to San Francisco, were photographed for this study.

Another investigator, William Hart of the U.S. Dept. of Agriculture, has used photographs brought back by the Conrad crew to isolate areas along the Texas-Mexico border where insect pests, including fire ants from the United States and fruit flies from Mexico, might cross to infect new areas.

Data for mapping projects in ten nations of the Western hemisphere were gathered on a number of earth resource passes, including several passes during which the space station remained in solar inertial attitude, with its sensors pointed obliquely at the earth. Studies of sea conditions, sedimentation and marine biology were supplied with information by Skylab's early flights. During the second mission, the millions of tons of seaweed in the Atlantic Ocean's Sargasso Sea were surveyed by the earth resource instruments.

The final Skylab flight, with a crew composed of space "rookies" Gerald Carr, Dr. Edward Gibson, and William Pogue, began November 16th and ran for eighty-four days. The astronauts returned to earth on schedule and in good physical condition February 8, 1974. One of the crew's most important assignments was to gather fresh information on the Comet Kohoutek. Scientists expect the nucleus of the comet is composed of primordial material about 4.6 billion years old—more ancient than anything found on the earth or the moon, which have been modified by cataclysmic events since formation of the solar system. A comprehensive study of Kohoutek thus should greatly enhance knowledge of the processes involved in the formation of planets and comets.

The final Skylab flight crew also concentrated on gathering scientific data on seasonal changes, the development of sea and lake ice, snow cover patterns, changes in vegetation in the northern and southern hemispheres, and major storms. Additionally, data were sought for many investigators conducting agricultural, forestry, urban and regional planning, and pollution studies—in which information must be acquired at several separate times during the year.

The astronaut's visual observations of sand dunes, volcanoes, ocean waves, cloud forms, geological features, and the like will not only support dozens of scientific projects underway on the ground, but they will also aid immensely in the future planning of researchers who may observe earth in the 1980s as passengers aboard the space shuttle

Summarizing the results to date from Skylab, Dr. Fletcher

said they have proved that man can live and work effectively in space for long periods. He called the program "one of the most significant scientific ventures of all time," and said the thousands of photographs and miles of recorded data brought back about man, the earth, and the sun will "overwhelm" scientists and keep them busy for five years.

Part Two

the
earth applications

O
Chapter 4
global communications

And the whole earth was of one language, and of one speech.

—GENESIS 11:1

It is said matter-of-factly now, with no longer even a hint of awe or wonder on the part of the broadcaster or the viewer: "This program is being brought to you live and in color via satellite." Historic visits of American officials to China and Russia . . . Winter and Summer Olympics from Europe, Asia, and North America . . . on-the-spot coverage of momentous news occurrences, as they develop, from anywhere on any continent . . . astronauts exploring the surface of the moon a quarter of a million miles away. . . . All these events, and hundreds of others, are beamed to television sets in homes around the world, routinely, by communications spacecraft positioned strategically above the earth.

Satellite television coverage of the Twentieth Olympic

53

Games in Munich, Germany, in August–September 1972 set a record for overseas telecasting of a special event. Four satellites were used in transmitting 1023 half-channel hours of TV coverage to ground stations in 33 countries. And it is believed that more people saw, live, the first American manned landing on the moon—Apollo 11 in July 1969—than had seen any other single event in the history of civilization. The audience was estimated to be more than 600 million.

And yet, although it affects the greatest number of people, television is not the primary beneficiary of the versatile spacecraft. In fact, TV represents only a small percentage of the workload of the international system now in operation above the planet. The biggest payoff is coming in direct economic gain to world commerce. Business efficiency has been immeasurably increased by the satellite's ability to provide cheaper and more reliable long-range communications, and through the ease of contact it affords by virtue of its extraordinary capacity.

For communications purposes, satellites offer unique advantages because they can overcome the "ground-bound" problems associated in spanning oceans and continents with submarine cables, land lines, and microwave radio stations for the long-distance transmission of radio, telephone, and television signals.

Because microwave travels in a straight line, for example, relay stations must be spaced reasonably close to each other—about every thirty-five miles or so—to allow for the earth's curvature so they can receive, amplify, and retransmit signals. To space stations this distance apart around the world would be prohibitively expensive. Similarly, submarine cables cost a lot to install and maintain, and generally have limited capacity.

One satellite, however, placed in a synchronous orbit 22,300 miles above earth at the equator, can remain in a fixed position and provide communications relay coverage across an ocean to two or more continents. Three such satellites spaced at 120-degree intervals around the equator can effectively cover the entire globe, except for small areas around the north and south poles. Thus, in placing satellites above the Atlantic, Pacific, and Indian Oceans, a truly worldwide communications network is established.

The satellites remain permanently in place because at this

synchronous altitude, their speed in orbit matches the rotational speed of the earth. They appear to hover over one spot.

The first transoceanic television transmission occurred only in 1962.

Historically, the idea of posting artificial earth satellites in orbit as relay stations for communications purposes is credited to space visionary and science fiction writer Arthur C. Clarke. His 1945 proposal for a stationary chain of synchronous satellites ringing the earth was before its time, however, and aroused little initial interest.

But by 1959, realizing the potential, NASA began a program to develop the technology necessary to establishing such a system. Three years later, on July 10, 1962, millions of television viewers in the United States and a few in France and England, watched a taped black-and-white picture of an American flag flapping in the New England breeze to the recorded accompaniment of "The Star-Spangled Banner." Picture and sound, transmitted skyward over the Atlantic from a huge horn-shaped antenna near Andover, Maine, were retransmitted to Andover and to Holmdel, New Jersey, from a new earth satellite—Telstar I, built by the American Telephone & Telegraph Company and launched by NASA fifteen hours earlier at Cape Canaveral.

Telstar signals also were picked up by ground stations in France and England. Two weeks later mass audiences on both sides of the ocean watched the first international exchange of live television. The era of Space Age communications—one that would dramatically change the lives of everyone living on earth —was born.

Telstar 2 soon followed, then came the more powerful Relay satellites 1 and 2 and three Syncoms—Synchronous Communication Satellites—whose speed in circular orbits 22,300 miles above earth was the same as the speed of earth in rotation, so that the Syncoms, "parked" at 120-degree intervals around the globe, could provide uninterrupted communications services for most of the world.

Among the memorable international telecasts these experimental spacecraft transmitted were the inauguration of Lyndon Johnson, the funerals of Winston Churchill, Pope John XXIII, and President John F. Kennedy, and the 1964 Olympics. The Telstars,

Relays, and Syncoms did much more than this, however. They developed the technology needed for the advent of commercial communications satellites.

President Kennedy, on August 31, 1962, signed Public Law 87–624—the Communications Satellite Act. "The ultimate result will be to encourage and facilitate world trade, education, entertainment, and many kinds of professional, political, and personal discourses which are essential to healthy human relationships and international understanding," Kennedy said upon signing.

Among other things, this act provided for the establishment of a new privately owned corporation to serve as America's entity in international satellite communications. Thus the groundwork was laid for the creation of the Communications Satellite Corporation, or Comsat.

By an Act of Congress in February 1963 Comsat was given a charter to establish a commercial system "by itself, or in conjunction with other countries, but using investor funds."

A year and a half later the International Telecommunications Satellite Consortium—Intelsat—was formed, initially with fourteen nations, as a joint venture to establish the global commercial communications satellite system. Today its more than eighty nation-members represent approximately 95 percent of the telecommunications traffic on earth.

Comsat acts as manager for the Consortium, each member of which shares in the costs and services of the satellite operation. The ground stations employed in the system are owned by the governments of or private organizations in the countries where they are located. Intelsat obtains its satellites, built by the Hughes Aircraft Company, El Segundo, California, through contracts with private industry. NASA is paid to launch them.

The first commercial communications satellite—Early Bird 1, or Intelsat 1—was launched on April 6, 1965, and successfully placed in synchronous orbit over the Atlantic Ocean.

"These satellites herald a new day in world communications," President Johnson said at the time. "For telephone, message data, and television new pathways in the sky are being developed. They are sky trails to progress in commerce, business, trade, and in relationships and understanding among peoples."

Early Bird 1 had a capacity for handling 240 telephone

conversations or one television program. Today's advanced generation of communications satellites—Intelsat 4s—are electronic marvels that can transmit:

- as many as 9000 one-way or 5000 to 6000 two-way phone calls; one of these satellites has the capacity to exceed the total of all the world's submarine cables;
- up to a dozen simultaneous color television programs from continent to continent, spanning the world's major oceans;
- any combination of communications traffic, including data and facsimile, or tens of thousands of teletype circuits.

Each Intelsat 4 satellite has a life expectancy of about seven years. They are big—about 18 feet tall and 8 feet in diameter, and weigh nearly 1600 pounds when positioned in orbit. Sophisticated electronics systems give them a communications handling capacity more than 35 times that of Early Bird 1.

Three Intelsat 4s, one each stationed over the Atlantic, Pacific, and Indian Oceans, provide global coverage. They are equipped with two four-foot-long "spot-beam" antennas which aim signal energy to earth like rays of a searchlight to concentrate thousands of voice channels and television broadcasts within the more densely populated communications centers of the world.

The most technically advanced satellite ever developed for space communications, Intelsat 4 has two main elements: a spinning rotor containing sensors and fuel systems to position and orient it in orbit, and a stationary platform that keeps the antenna system constantly aimed at earth.

The spot-beam antennas, which are movable, can be aimed to cover other areas within view of the satellite, such as South America or Africa. Their signal energy is concentrated in a 2000-mile circle on earth; however, an effective radiation power of up to 3000 watts per spot-beam channel enables more ground stations than ever before to use the satellite.

Without these satellites there could be no global television, owing to the limiting effect of the curvature of the earth as previously mentioned, on the range of direct radio and TV transmissions. A single TV channel is roughly equivalent to 1000 voice channels, and provision of underwater cable capacity for TV would have been prohibitively costly.

Also, before the advent of the Intelsat system, only 500 cable circuits were available for transoceanic phone conversations. Today, Intelsat alone uses more than 4000 circuits. In the few years since Early Bird, in 1965, commercial traffic has grown to more than 2500 hours of television a year, and more than 4000 full-time leased two-way voice circuits.

The three major types of "traffic" are voice communications, data transmission, and television.

With their low-cost circuits, communications satellites have had a dramatic impact on transoceanic telephone charges. (About three and a half million ocean-spanning calls were made in 1960 and the figures projected for 1980 run anywhere from 40 to 100 million. One of every two such calls today is made via satellite.) A three-minute station-to-station phone call from New York to London, for example has been lowered 40 percent, from $9 to $5.40. Monthly rental charges for the American half of a leased voice-grade circuit between New York and Paris have been reduced from $10,000 to $4750, or by more than 50 percent.

An Intelsat 4 stationed over the Atlantic provides direct high-quality telephone circuits between earth stations serving such distant cities as New York, London, Paris, Munich, Rome, Madrid, Teheran, Rabat, Beirut, San Juan, Mexico City, Panama City, Rio de Janeiro, Buenos Aires, Santiago, Lima, Bogotá, and Caracas. Similar capability exists in the Pacific and Indian Ocean regions, where countries with earth stations are also linked via satellite to all other countries having stations.

The broad-band transmission capability of the satellite system also offers service potentials never before available. A unique capability is simultaneous broadcast of voice and data. Another is the possibility of international picture-telephone service. And the multipoint versatility of satellites permits simultaneous transmission to several countries.

Before Intelsat, direct overseas dialing was theoretically feasible between only a few points. The growing number of earth stations around the globe, however, is quickly extending direct-dial capability worldwide.

A specially allocated bank of frequencies in today's communications satellites makes transoceanic telephoning possible on

Following final inspection, an Intelsat 4 communications satellite is prepared for mating with its launch vehicle at Cape Kennedy.

a demand assignment, or as-needed basis. This has particular significance for the many emerging or developing nations whose traffic requirements don't justify preassigned permanent circuits. Through demand assignment they can communicate directly with other countries without having to establish permanent full-time facilities.

There are numerous potential applications other than the public telephone network. Communication of voice and data to ships at sea and aircraft in flight, and other uses requiring flexible communications between varying points on the earth's surface are all possible via satellite.

One of the fastest growing areas of communications satellite use is data transmission. Intelsat offers a new and efficient way to send data internationally because it uses bandwidths and transmission rates not economically practical in conventional cable systems.

The unique advantages of the satellite system are its ability to interconnect any two points in the world and provide data transmission service equal to or exceeding the speed and accuracy standards of ground systems.

Intelsat offers tremendous potential for new uses of television in education, science, industry, medicine, and government. Today practically every heavily populated area on earth is served by satellite communications through more than eighty antennas at some seventy earth stations in over fifty countries.

TV by satellite has experienced an incredible growth rate— from an average of about five transmissions a month in 1965 to more than a hundred a month in late 1973. And the programs are reaching ever increasing audiences. The use of transportable earth stations and "two-hop" satellite relays has further extended television's reach, enabling huge audiences around the world to witness live transmissions of memorable news events. Coverage originating in Japan, for instance, can be beamed to a satellite over the Pacific, bounced down to United States ground stations, relayed to an Atlantic satellite, and flashed to viewers in Europe. Vast distances, difficult mountainous or jungle terrain, and widely dispersed islands can be spanned easily by use of a satellite communications system. This offers developing countries the opportunity to improve their communications in a quantum leap, with important economic advantages as well as such social benefits as educational television.

The feasibility of a worldwide medical diagnostic center has been demonstrated by transmitting electrocardiograms and other data between the United States and Europe by satellite, thereby establishing the potential of making vital information available wherever emergency medical care is needed.

The use of space has also expanded the horizons of closed-circuit television. A variety of such telecasts has demonstrated a large potential for commercial and educational application in the financial, industrial, scientific, educational, and governmental communities.

In business, for example, a major raw materials producer dramatized the initial shipment of iron ore from its new multi-billion-ton reserves in northwest Australia. Via satellite the company relayed a telecast of dedication ceremonies from Sydney to Tokyo, New York, and London, where potential customers—world steel producers—had been assembled in conferences as part of a simultaneous and coordinated worldwide sales effort.

Companies with international interests use closed-circuit satellite television to enable audiences of shareholders on different continents to view, live, their annual meetings.

More than a thousand physicians and medical scientists assembled in several European countries to participate in an international closed-circuit TV conference on aerospace medicine and cancer and tuberculosis research. The three-hour program originated in Houston, Texas, and was viewed at the same time in ten cities in the Federal Republic of Germany, Austria, and Switzerland.

In an earlier demonstration, physicians in Europe watched a pioneering open-heart surgical procedure live via transatlantic closed-circuit TV and were able to ask the operating team questions on technique at the same time.

The closed-circuit television potential of communications satellites is virtually limitless. One of the major factors contributing to the rapid growth in the use of satellite television has been the substantial reduction in charges since the introduction of commercial service. The present charges for color TV service from the United States to Europe—about $2500 for the first ten minutes—is 77 percent less than the peak charge for the same service in 1965.

The era of domestic communications satellites for use by individual nations or groups of nations was advanced November 9, 1972, with the launch of Anik 1 by NASA for the Canadian government. The official organization, Telesat Canada, was established by Act of Parliament on September 1, 1969, to own and operate Canada's domestic satellite communications system.

On April 20, 1973, a second Anik was launched into synchronous orbit about 22,500 miles above Canada. These satellites are somewhat smaller than the Intelsat 4 series. Nevertheless, each Anik has the capability to relay ten color TV channels, or up to 9600 telephone circuits, or any "mix" desired. They also have an in-orbit life expectancy of seven years.

With them constantly covering Canada—the world's second largest country in area—there is for the first time in Canadian history accessibility for almost everyone to transcontinental communication. Prior to the age of Anik, the only contact hundreds of small communities, especially in the thinly populated northern regions, had with populous areas was by shortwave radio. And this is subject to severe interference from weather and other factors. Anik provides reliable inexpensive two-way communications for everyone in Canada near a receiving station. Such stations have been positioned throughout the nation's ten provinces—from large "heavy route" ones in Vancouver and Toronto, to smaller, simpler stations at such distant points as Baffin Island or Igloolik, off the Melville Peninsula. These, incidentally, provide telephone service to isolated arctic communities.

Additionally, remote TV stations have been set up at dozens of other locations that previously had not had access to live television. Through these and the satellites TV has become available in most of the inhabited areas of Canada—as far north at least as Frobisher Bay and Resolute Bay.

The total effect of the domestic satellite system has been to provide a large increase in circuits available in the populated Canadian south, and to put the more distant communities into better contact with the rest of the nation. Without Anik there would have been no practical way to accomplish this.

Comsat began studies of potential users of domestic satellites in the United States several years ago. Business concerns with the most direct interest include broadcasters, news wire services, communications common carriers, newspapers, computer service companies, Western Union, airlines, and community-antenna television companies.

The Federal Communications Commission (FCC) has responsibility for regulating the use and control of domestic satellite systems. "The combination of computers and communications using satellites could have a significant impact on all our lives," says Bernard Strassburg, chief of the FCC's Common Carrier Bureau. "Our objective has been to establish a climate—to provide the opportunity and incentives so people will risk their capital in an effort to develop new public benefits."

The first American domestic satellite communications system began service in 1974. It is the Westar System of Western Union, which uses Anik type satellites and technology. Westar 1 was launched into orbit April 13, 1974. Under an FCC ruling Comstat was directed to form a subsidiary to conduct all of its domestic satellite activities. That subsidiary, Comstat General, has signed an agreement to lease three satellites in orbit, and to provide a spare on the ground, to the American Telephone and Telegraph Company.

Each of these satellites, the first to be launched in 1975, will have a capacity of approximately 14,400 two-way telephone circuits, which will add diversity and flexibility to AT&T's nationwide switched network. Earth stations for communications service with the satellites will be provided, owned, and operated by AT&T.

As a step toward implementing another domestic satellite program in the United States, Comsat General has joined with MCI Communications Corporation and Lockheed Aircraft Corporation in a separate company to provide multipurpose domestic United States satellite services to all customers, except AT&T.

Additionally, the FCC granted authority early in 1973 for an interim domestic communications satellite system through the leasing of Canada's Anik satellites. This grant was to the American Satellite Corporation, a joint venture of Fairchild Industries.

The first relay of a telecast across the United States by domestic communications satellite was in June 1973. The cable

TV program featured Carl Albert, Speaker of the House of Representatives, and originated in Germantown, Maryland. It was transmitted by an Amsat antenna via Anik, to Anaheim, California, where it was received by a new mobile ground receiving station built for TelePrompter Corporation, and delivered to cable television systems in Long Beach and Newport Beach.

Temporary United States service via American Satellite Corporation began in the fall of 1973. The company's own domestic satellite system is expected to be operational late in 1974. Just how many different systems eventually will be placed in service in the United States is, at this time, difficult even to estimate.

The American domestic communications satellites will provide the same range of services as the current Intelsats, and generally will be used for transcontinental relay of voice communications, data transmission, and television programming. American Satellite Corporation officials have proposed a rate structure that will enable customer-users to cut communications operations costs by up to 50 percent. The cost for full-duplex voice channel via satellite per month from New York to Los Angeles will be about $1200. The present private line base rate per month is $2400.

Across the Atlantic the European Space Research Organization (ESRO) is far along on planning for a domestic communications satellite system that can be used throughout Europe. The hope is to have it operational by the late 1970s or early 1980s.

The spacecraft for all national domestic communications programs are launched by NASA on a cost-reimbursable basis. In the immediate years ahead other satellite systems will be designed, developed, and operated by other nations and groups of nations around the world.

Today Intelsat 4 satellites, positioned above the Atlantic, Pacific, and Indian oceans, provide truly global communications coverage. But as sophisticated and versatile as these spacecraft are, they will be succeeded in the 1970s by even more advanced systems. Two improved Intelsat 4A satellites are planned for launch in 1975. Each will offer approximately double the communications capacity of their counterparts in the current network. By the time they are in operation, it is expected that more than one hundred antennas in sixty countries will be linked by satellite communications.

Already under development and planned for launch in the 1976–1978 time period, Intelsat 5 is expected to weigh more than 3000 pounds, and to have an operating life of ten years and a communications capacity five times greater than that of Intelsat 4.

Meanwhile, NASA is continuing experimental programs aimed at further advancing the art of communicating through space. One such project is called ATS—for Applications Technology Satellites. New systems and concepts are tried on these craft. Since June 1970, for example, NASA has been cooperating with Alaska in a variety of investigations using the first satellite in this series, ATS-1. Some of the most promising studies involve education. In one instance, two dozen communities participated in a series of public radio broadcasts. At these remote sites teachers conferred by two-way voice communications via satellite with specialists at the University of Alaska to upgrade their teaching skills.

In February 1971 the Pan Pacific Education and Communication Experiments by Satellite (Peacesat) program was initiated at the University of Hawaii. Inexpensive, mobile ground stations were set up at the University's Manoa and Hilo campuses, the University of the South Pacific in Fiji, and the Polytechnic Institute at Wellington, New Zealand. Experiments relayed to this network through ATS-1 have proven the feasibility of holding classes and seminars, sharing library facilities, and coordinating administrative projects via satellite. Additional ground stations were set up in New Guinea, American Samoa, Saipan, and Truk. For some of these locations the network provides the first reliable instantaneous communication with world population centers. Such a system could serve as a cultural clearing-house, linking the peoples of the Pacific to build greater understanding, unity, and progress.

In the years ahead there are likely to be universities of the air which will permit many students who could not otherwise attend a college to obtain degrees while spending only a small percentage of their time in academic residence.

"The role of communications [satellites] is not limited to

commercial use," said President Lyndon Johnson in 1968. "It must also provide a 'network for knowledge' so that all peoples can share the scientific, educational and cultural advances of this planet. Failure to reach these goals can only contribute to apathy, ignorance, poverty, and despair in a very large part of the world. Success in our telecommunications policies can be a critical link in our search for the understanding and tolerance from which peace springs. Communication by satellite is a tool—one of the most promising which mankind has had thus far—to attain this end. We must use it wisely and well."

Another potential application of communications satellite systems is in the field of law enforcement. Before his death in March 1972, J. Edgard Hoover, director of the Federal Bureau of Investigation, requested NASA's assistance in a communications systems design study possibly involving satellites. The FBI operates a nationwide computer/communications system known as the National Crime Information Center (NCIC). It is used by local, state, and federal law enforcement agencies through more than one hundred control terminals in all fifty states and Canada. NCIC contains almost 2.5 million active records on wanted persons, identifiable stolen property, and stolen vehicles. The system currently handles a daily average of over 56,000 transactions. The Bureau is developing a computerized criminal history file for use in the NCIC system which will require a substantial increase in network capacity. Requests for this information, from agencies across the continent, are greatly increasing too. Satellites may prove to be the most efficient means of communications for speeding such data to the control terminals.

In a separate NASA study proposed by the Justice Department in 1971, NASA cooperated with California and Florida law enforcement agencies to determine ways in which satellite communications could speed the judicial and criminal identification process by making such records as fingerprints, suspect descriptions, and photographs more rapidly available to courts and agencies across the nation. During this trial run with an ATS satellite, fingerprints transmitted at various data-rates, including some at a far greater speed than possible with telephone circuits, proved to be of uniformly high quality. Although these transmissions were experimental, real criminal prints were used and three criminal identifications actually were made over the satellite link. There are a number of other promising areas for the communications satellites of the 1970s and beyond. Postal service efficiency, for one, will improve with the advent of electronic mail handling. Satellite links may be expected to carry a majority of the long-distance letter mail, making inexpensive overnight delivery possible to all but the most remote locations. Messages would be sent from one post office to another via satellite and be delivered in the regular mail.

New satellite systems for aviation and maritime traffic management will reduce the accident rate and increase the effectiveness of transoceanic and coastal transportation of goods and people.

"In the relatively short span of seven years," President Nixon said in 1972, "communication by satellite has changed the world forever. We now live, in one very real sense, much closer to other peoples and to faraway events. The fast-developing science of satellite communications must rate as one of the true marvels of the twentieth century—a technological triumph that is bringing greater understanding to a world badly in need of closer ties and deeper insights. . . . Just as this technology has enabled men to speak to each other across the boundary of outer space, so, I am convinced, satellite communications will in future years help men to understand one another better across boundaries of political, linguistic, and social nature. World peace and understanding are goals worthy of this new and exciting means of communication."

But perhaps author-prophet Arthur C. Clarke, who conceived the idea of relaying information around the world through spacecraft in stationary orbit three decades ago, summarized the ultimate value of such a system as well as anyone:

"Communications satellites will end ages of isolation, making us all members of a single family, teaching us to read and speak, however imperfectly, a single language," he said. "Thanks to a few tons of electronic gear 20,000 miles above the equator, ours will be the last century of the savage—for all mankind, the Stone Age will be over."

○
Chapter 5

from stars to satellites

One of the least developed areas of applications satellites to date, yet one that offers tremendous potential in the 1970s, is navigation, which is closely linked with communications.

"Future navigation systems will provide accurate position determinations for both aircraft and ships with a high degree of accuracy (one mile or better within the not too distant future)," says United States Congressman Joseph Karth of Minnesota.

"Stretching our imagination into the future," Karth adds, "the Maritime Administration and industry people in the shipping field have been looking to the day of the automated ship. This is a ship that will leave a port in the United States with a crew consisting of a handful of men, primarily electronics specialists who will operate and maintain the computers and other specialized equipment. The satellite position information would be fed directly into ship's computers. The ship would stay on a perfect prescribed course to its final destination.

"Of prime importance, too, is air traffic control and maritime coordination. Collision avoidance can be greatly improved by the use of such a combination of communications and navigation systems."

The idea for a navigation satellite system actually can be traced back more than a century. Again, it was an imaginative author who first conceived it. Edward Everett Hale, best known for his story "The Man Without a Country," envisioned a large artificial satellite circling the earth in an orbit over the poles and passing along the Greenwich meridian. He described the concept in his book *The Brick Moon*. Hale reasoned that ships at sea could take bearings on this man-made moon and thus fix their positions more accurately.

Ninety years later, in 1959, the United States Navy launched Transit 1A, the first real navigational satellite, to help guide its submarines. Several more spacecraft were orbited over the next few years, and by 1965, both the United States and the Soviet Union had operational systems—used primarily for the benefit of subs and a few of the larger surface ships.

(Before the advent of the "navsat," the Navy frequently experienced navigational errors of two to three miles in good weather and as much as fifty miles in bad weather in pinpointing the location of fleet ballistic missile submarines. The network of navigational satellites makes possible position "fixes" with the errors as small as the length of a submarine.)

The classical concept of an orbital navigation system depends upon accurately knowing the satellite's position and then finding one's position, be it in an aircraft or ship, relative to it. In other words, the satellite becomes a known landmark: the only one visible on the broad oceans. Stars are used in the same manner in stellar navigation, but they are not always visible. Also, stellar fixes are too slow and difficult for aircraft flying near or past the speed of sound.

The advantage of navigation satellites is that signals can be automatically received and analyzed by computers, giving pilots their positions rapidly and continuously. They operate under all weather conditions and can supply position information to ships anywhere.

The Navy system uses a number of satellites in circular near-polar orbits about six hundred miles above earth. Twice a day new orbits are calculated by tracking stations for each of the satellites, and these data are stored in the spacecraft's memory unit. The satellite then transmits the orbital data at two-minute intervals.

Navigators on submarines or surface ships determine when the satellite is directly overhead by analyzing its radio signals, which change frequency as the satellite approaches, then recedes from, the vessel. This information, together with the orbital data, the time the signal was received, and the speed of the ship, is fed into a computer aboard the ship which rapidly calculates and fixes the ship's exact position.

Navigational satellites have been used almost exclusively by the military, for one simple reason: they are still too expensive for commercial application. The cost of the shipborne computer and associated communications equipment is too great for all but a few specialized craft devoted to such pursuits as oceanographic research and oil exploration. Such sophisticated equipment is also costly to operate and maintain.

However, as the state of the art advances and newer, less complex systems are designed and tested, the possibility of a truly global navigation satellite system, for the use of nations around the world, looms closer and closer to reality.

Recent Transit launches have carried specific experiments designed to lower costs for users. A disturbance compensation system (DISCOS) has been added. Its purpose is to keep a satellite in a precise orbit, so precise that reference tables can be published listing its exact path for years into the future.

Once this can be made operational, satellites could orbit in fixed preassigned paths. Shippers would no longer need costly computers to find out exactly where the satellites were. Even fishing and pleasure boats with radio receivers and charts would be able to find their bearings.

Although NASA is not building a navigation satellite, it is continuing to conduct a number of Navy experiments on ATS and Nimbus spacecraft. On one, the Omega Position Location Experiment, ships and aircraft used the ATS 3 craft to fix their locations within three and five miles, respectively. Similarly, a jet aircraft fixed its position to within four miles by simultaneous radio ranging measurements on ATS 1 and 3.

The Navy's Fleet Weather Facility (FWF) reports that satellite imagery provided by NASA-developed meteorological satellites is becoming "indispensable" to shipping operations in the Arctic and Antarctic. In fact, Navy weathermen predict that as a result of satellite pictures showing the location of ice masses, shipping operations may be extended several months—perhaps ultimately through the whole six-month polar night.

The navigation season is already being extended a number of weeks during the polar darkness periods because of the cloud-piercing capabilities of the microwave sensors on NASA's Nimbus satellite, which is being used operationally by the FWF.

"Satellite imagery that sees the ice pack at both poles day and night and even through clouds with Nimbus 5's new microwave radiometers, has not only extended the navigation season, but has shown us more ice than we ever knew existed," says Lt. Cmdr. William Dehn, head of FWF's Sea Ice Department.

The Navy can now chart all of Antarctica in winter, even through the almost constant cloud cover. The images show the continent's ice cover, its shape, and major sea ice features.

In cooperation with the Maritime Administration (MARAD) of the Department of Commerce, NASA has been exploring ways in which satellite systems providing communications with and position fixing of oceanic and coastal ships could contribute to a safer and more efficient American shipping fleet.

For one thing, a navsat could double as a communications link to vessels at sea, permitting continuous voice contact between ship and shore. This would allow the transmission of up-to-the-minute regional weather advisories so that ships could steer clear of storms.

Also, precision navigation is in itself a practical economic tool in that it can save fuel costs and reduce time at sea. Direct contact between fleet shipping headquarters and vessels on the seas opens the possibilities of wider flexibility in scheduling and routing.

But the real benefit of an operational navsat system is in

human safety—not only in collision avoidance, but in post-accident rescue. Unfortunately, all too often search and rescue craft experience delays or total failure in their efforts to find a downed aircraft or a distressed ship. Frequently, the last position reported turns out to be miles from the real location. But through continuous monitoring from space the traffic-control centers would know the precise position of any craft in trouble, eliminating the search period during which every minute may literally mean life-or-death.

Closely akin to the navigation and the communications families of spacecraft is the air-traffic-control satellite, or Aerosat. This system differs from the others in that it centralizes the control of aircraft positions relative to one another rather than determining their absolute geographical positions.

NASA has been working with the European Space Research Organization (ESRO) to explore the possibility of cooperative air-traffic-control experiments on satellites. Consequently, NASA–ESRO, Canada, and the United States Department of Transportation will participate in air-traffic-control and maritime satellite tests to be run on an ATS spacecraft beginning in mid-1974.

Aerosats could provide operational communications, such as airline message traffic, navigation, traffic control, collision avoidance, passenger telephone service, weather advisories, and search and rescue data. They could also be used by ships, eliminating the need for two separate systems. Space experts believe satellites of this type will eventually be able to establish the location of aircraft or ships within a positional accuracy of sixty feet or less, and will be able to operate continuously, regardless of weather conditions. Based on current technology and budgetary restraints, however, it is not likely that an Aerosat system will be fully deployed until sometime in the 1980s.

Whatever the date, new satellite systems for aviation and maritime traffic management will substantially reduce the accident rate and increase the efficiency of transoceanic and coastal transportation of goods and people.

Despite the modernization of transportation systems around the world, particularly the airlines, there is much room for improvement. Because of navigational shortcomings, for example, air traffic control regulations now demand a great lateral separation of planes—up to 120 miles—as an anticollision measure. Thus, when a number of aircraft depart from a terminal within minutes of each other, only one can take the direct shortest-distance-between-two-points route. The others, to comply with safety regulations, must follow wide-sweeping flight paths that lengthen travel time and increase fuel expenditure. Aviation experts have estimated such extra costs may run as high as $30,000 or more per year per airplane, which multiplies to a very substantial figure for airlines operating large fleets.

An air-traffic-control satellite system using spaceborne navigation techniques, however, could reduce the required lateral separation to thirty miles or less. It has been estimated that the dollar savings for jets flying the North Atlantic alone could be close to $20 million a year.

Also related to navigation is the science of geodesy, which satellites are projecting into an era of unprecedented growth and accuracy. Basically, geodesy refers to the mathematical determinations of the earth's size, shape, and mass; variations in earth's gravity; and distances between and locations of points on earth.

The United States National Geodetic Satellite Program is designed to reap new knowledge about our planet's structure, origin, and history. Early geodetic satellites were the first to find that earth was slightly pear-shaped rather than round, and that the equator is elliptical instead of circular.

Further, sensitive instrumentation aboard these orbiting spacecraft has found that some small islands in the middle of the oceans were actually dozens of miles from where maps placed them. Through the use of satellites, geodesists can now calculate distances between points thousands of miles apart to within several hundred feet of the actual distance.

Among the principal objectives of geodetic satellites are the establishment of a single common worldwide geodetic reference system, and the improvement of global maps to an accuracy of about thirty feet. The United States Department of the Interior estimates that up-to-date topographical maps would be worth approximately $700 million a year to industry in the United States alone. Worldwide, the figure would undoubtedly be several billions of dollars.

NASA plans to orbit a Geodetic Earth Orbiting Satellite—

GEOS-C—late in 1974. Among its assigned tasks are to refine the description of the earth's gravity field; apply satellite radar altimetry and satellite-to-satellite tracking techniques to investigations in earth geophysics and oceanography; develop an earth reference system by the precise location of approximately eighty-six widely separated control points on the surface of the globe to an accuracy of plus or minus ten meters; and to demonstrate the feasibility of using a satellite radar altimeter to determine the geometry of the ocean surface to an overall accuracy of five meters.

Through the 1970s and into the early 1980s NASA planned five new applications spacecraft to monitor the physics of the earth and its oceans.

- Scheduled for launch in 1976, Lageos will be a small, dense, passive satellite covered with laser retroreflectors intended to provide a permanent reference point for precision earth-motion measurements such as continental drift. It is to have an orbital lifetime of at least fifty years.
- Seasat-1 is to be the first experimental ocean dynamics monitor for mapping the topography of the ocean surface by means of satellite altimetry. It will record sea-state, ocean currents and circulation, and such transient phenomena as tsunamis (tidal waves), storm surges, and barometric effects. Such data are vital to commercial interests, because it is estimated that global real-time coverage and forecasting of precise ocean-surface conditions, using direct communications, could save the equivalent of twelve days, or more than $50,000 per ship per year. This would amount to an approximate half-billion dollars per year for 10,000 cargo ships and tankers. Research and development on this satellite program will extend through the mid-1970s.
- Seasat-2 would be the prototype of an operational global ocean dynamics monitoring system, and could be in service in the early 1980s.
- Geopause is a planned precision tracking satellite for extreme accuracy measurements required to support gravity-field and ocean-dynamics satellite missions. The experimental phase of this program is scheduled to extend from late 1975 into 1979.
- Gravsat is a low-altitude satellite for measuring the fine structure of earth's gravity field. Its research-and-development schedule is similar to that of Geopause.

The overall economic boon such satellite systems promise is immense. Charles W. Mathews, associate NASA administrator for applications, asserts that "the maritime shipping industry could save hundreds of millions of dollars a year through optimum-time routing of ships, improved scheduling of ship arrivals and unloading, more efficient use of personnel and equipment, etc., if they could improve their ability to route, schedule, and predict ship arrival times.

"The availability of an operational Ocean Dynamics Monitor System could also create a new industry to supply ships and shipping-traffic-control centers with equipment for receiving and displaying maps of observed and predicted ocean-surface conditions."

○
Chapter 6

worldwide weather watch

Early in June 1972 pictures relayed to ground stations from a satellite in earth orbit recorded the birth and development of a storm disturbance off the Yucatán Peninsula. To concerned meteorologists at the Department of Commerce's National Oceanic and Atmospheric Administration (NOAA), the pictures revealed that the young storm was acquiring an unusually large envelope of air circulation.

On June 14, NOAA issued an initial satellite weather bulletin indicating a high degree of probability that tropical storm development was imminent. As the disturbance continued building and moving toward the northwest Florida coast, across the Gulf of Mexico, the satellite doggedly tracked its progress and weather bulletins containing location and intensity information were issued daily.

The storm, which reached hurricane force by June 17, was named Agnes. Over the next several days it struck northeastward across Georgia and South Carolina, then moved out to sea along the eastern seaboard and backtracked inland over Pennsylvania, New York, and the Great Lakes region.

Agnes caused severe flooding, especially in Pennsylvania and Virginia, and property damage ran into the billions of dollars. But because its development and path had been so closely watched from space and because hurricane warnings had been flashed ahead to residents of the areas involved, the loss of life from one of the most violent and destructive tropical storms of the twentieth century was held to an incredible minimum. Fewer than a hundred people were killed, and most of these deaths were associated with the flooding that occurred after Agnes had passed.

Before the Space Age and the subsequent orbiting of unmanned meteorological satellites, storms were often born unobserved in the tropical seas and swept undetected across islands and into coastal areas bordering the Atlantic and Gulf of Mexico. Without proper warning, over five thousand unsuspecting people were killed shortly after the turn of the century when a vicious hurricane slammed into Texas. Another four thousand lost their lives in the West Indies when a killer storm swept through the islands. As recently as 1959, fifteen hundred people died as a result of a hurricane that wrought devastation across Mexico.

The inestimable value of weather observations and warnings from space was perhaps most dramatically emphasized on the night of August 17, 1969, when Hurricane Camille—the most intense storm to hit North America in modern times—lashed into the Louisiana-Mississippi coastline. Fierce winds of more than two hundred miles per hour and walls of water, surging twenty-four feet above mean sea level, flattened entire towns along the low-lying coastal region. Property damage approached $1.5 billion, an all-time record for a single storm, and tens of thousands of people were left homeless.

Yet, incredibly, only about 320 people lost their lives, and

Eerie, swirling eye of Hurricane Ginger was photographed from space by a Nimbus meteorological satellite. Tracking of such storms from space has helped to reduce loss of life and property through early warnings on earth.

fully one-third of these, again, died in the ensuing floods. Had it not been for the prompt warnings from the Weather Service, made possible by satellite and weather aircraft reports, some 50,000 of the 75,000 people who evacuated the area in the path of Camille might have been killed, according to government estimates.

Dr. Walter Orr Roberts, president of the University Corporation of Atmospheric Research in Boulder, Colorado, an organization supported by the National Science Foundation, has estimated that weather hazards cost the United States twelve hundred lives and $11 billion in property damage annually.

While the life-saving capabilities of spaceborne meteorological systems may well be their greatest boon to mankind to date, they are by no means the only beneficial contributions. In fact, even by conservative estimates the economic dividends to be gained through such systems—by more reliable, accurate long-range weather forecasting—have been placed at more than $2.5 billion annually in the United States alone!

At least $1 billion of this can be saved by the construction industry, where reliable weather information can mean optimum scheduling of work forces and protection of materials from damage by heavy rains or freezing temperatures.

It is projected that another $950 million a year can be saved in agriculture. More accurate long-range forecasting can help farmers stem losses by avoiding ill-timed irrigation of multiple spring plantings necessitated by excessive rains or unexpected droughts. This can directly result in more efficient vegetable, fruit, and even livestock production.

Two specific examples: In California raisin growers, alert for damaging rain, were assured, on the basis of satellite pictures, that such rain was not coming, thereby saving them thousands of dollars that would have been needlessly spent protecting their crops. In October 1972 Montana cattlemen, warned by photographs transmitted from earth orbit, took extra precautions to protect their herds against an unseasonably heavy snowfall that came the next day. The list goes on.

In transportation, meteorological information from space can effect rerouting or rescheduling air, land, and ocean traffic around delay-causing storms. Flight plans can be modified to assure passengers of safer trips. As just one example, the reduction in fuel and manpower costs of only one percent for cargo ship lines could produce a yearly saving of $150 million by 1975.

The National Academy of Sciences has estimated the dollar loss from weather damage to world shipping at $150 million annually. Flood- and storm-damage prevention, aided by accurate and longer-range forecasts, could contribute another $70 to $140 million in annual yield. The same information could aid in more efficient management of electric power generation, resulting in an additional $500 million annual saving. And these are merely some of the more obvious areas in which improved systems of weather information can be directly equated with economical and commercial gain. These examples do not include the international economic—and good will—benefits that accrue to the United States weather satellite program.

Satellites laden with sensitive instruments and unblinking electronic "eyes" scan the entire globe, continually relaying to ground stations around the world thousands of up-to-the-minute pictures of weather conditions as they develop. Such craft, from several hundred miles above earth, tirelessly make passing sweeps of the planet in less than two hours, covering the North Pole, crossing south across the continents and oceans until they are over Antarctica, and then heading north again.

Special TV camera systems catch the glint of oceans and mountainous ice fields, the billowing white expanse of clouds, and the familiar outlines of the world's land masses. Radiometers aboard the satellites detect heat and light radiation from earth, cloud, and sea. Every so often all the pictures and data stored during these continuous passes are relayed from space to ground stations, where, through complex computer networks, the information is speedily assessed, analyzed, and flashed across the United States and around the world. This enables meteorologists today to make the fastest, most accurate weather forecasts ever.

Man, of course, has always been aware of and concerned about the weather. Everybody talks about the weather, but nobody does anything about it, Mark Twain is said to have said, and for countless centuries this was literally true.

It wasn't until the sixteenth century, when Leonardo da Vinci constructed an improved wind vane, and the seventeenth century, when Torricelli invented the barometer, that man began to apply the principles and disciplines of science to meteorology. Even then, progress was painfully slow.

All this was drastically changed at 6:40 on the morning of April 1, 1960, when a small 300-pound hatbox-shaped spacecraft named Tiros 1 (Television and Infra Red Observation Satellite) was launched by NASA into earth orbit from Cape Canaveral, Florida. During its two-and-a-half-month active lifetime circling the world, it sent back to earth 22,952 cloud photographs. The Tiros program ran for more than five years—the last one was launched in July 1965—and included ten vehicles built by the Radio Corporation of America (RCA).

This prolific family of spacecraft contributed approximately a half-million usable weather pictures to help identify and track ninety-three typhoons and thirty hurricanes. Pictures from Tiros 3 were taken in time to make possible the largest mass evacuation in the history of the United States: more than 350,000 people were able to flee the devastating path of Hurricane Carla in September 1963, holding the loss of human life to a remarkable minimum.

Through the Tiros program more than 20,000 cloud patterns were analyzed, leading to the broadcast of 2500 important storm warnings worldwide. Tiros also proved, through orbital flight test, a revolutionary new device—the Automatic Picture Transmission (APT) camera system. Aboard the satellites APT relayed photographs of local cloud cover to ground stations within radio range. The comparatively inexpensive receiving equipment used with this system first encouraged the general use of satellite data in the preparation of local weather forecasts, and pointed up the immense international potential of an operational network of meteorological satellites.

Today, hundreds of weather stations in nearly a hundred countries and territories all over the world, including a number of the developing nations, have their own APT receivers. Signals flashed from the satellites are transcribed on paper by electric needles, which make six hundred passes in minutes—each pass corresponding to a scan made by the TV camera in space. These lines blend to form a continuous picture, as on a television screen.

NASA launched the first experimental Nimbus (Latin for "rain cloud") spacecraft on August 28, 1964. Larger, heavier at 1000 pounds, and more sophisticated than Tiros, through the late 1960s and early 1970s Nimbus became an orbital test bed for advanced meteorological instruments, cameras, and systems. Nimbus 1 transmitted some 27,000 photos of cloud-cover patterns in only twenty-four days. Later satellites in the series had the fantastic capability of sending to earth 667 million separate bits of information about global weather conditions a day! Nimbus television cameras and other instruments are capable of producing pictures of weather patterns in sunlight and nighttime portions of the satellite's orbit, a major technological breakthrough.

Over the years, Nimbus, originally conceived as a metero-

Weather analysis and forecasting have improved immensely over the past few years, largely because scientists can prepare meteorological maps from actual photographs taken by satellites in space.

logical project concerned primarily with providing atmospheric data for improved weather forecasting, has matured to become the nation's principal satellite program for earth-sensing research. Each succeeding spacecraft has grown significantly in sophistication, complexity, weight, capability, and performance. Data produced by the sensing instruments on Nimbus 1 through Nimbus 4 are now being studied and applied not only in meteorology, but also in oceanography, hydrology, geology, geography, cartography, and geomorphology (study of the earth's form, its general surface configuration, distribution of land and water, and the evolution of life forms).

The Nimbus 5 satellite is filling a critical gap in man's understanding of his environment by measuring daily the distribution of rainfall over the oceans that cover 70 percent of earth's surface. Measurements of water that evaporates from the oceans, changes again to water, and falls back to the surface are made by an electrically scanning microwave radiometer aboard the satellite.

Previously, weathermen had no adequate way to monitor ocean rainfall on a global scale. Knowledge of its extent and rate, according to meteorologist John Theon of NASA's Goddard Space Flight Center, can "give us a good handle on how much energy [heat] is being released into the atmosphere." In turn, he says, knowing rainfall rate and the amount of heat thus released will aid immensely in reaching the goal of long-range weather forecasts as well as improving short-term forecasts of severe phenomena such as hurricanes.

Although as spacecraft Tiros and Nimbus differ greatly, their contributions as experimental test-bed spacecraft for the advancement of spaceborne meteorological and earth observation programs have been complementary and basic to the development of the advanced generations of operational weather satellites that have followed them.

The first of these was established in February 1966 with the launchings of two Tiros Operational Satellites (TOS) by NASA for the Department of Commerce's Environmental Science Services Administration (ESSA). These craft and their successors, over the next four years, provided regular and continuous photographic cloud coverage of the entire sunlit portion of the world at least once a day from their orbital vantage points averaging more than 850 miles above earth. The torrent of thousands of pictures of spiraling cloud systems taken and transmitted to the ground daily by these satellites enabled meteorologists to forecast weather on earth two to three days in advance with unprecedented accuracy.

As the 1970s began a new series of advanced generation spacecraft took over the weather watch from space. These are Improved Tiros Operational Satellites (ITOS), and they are substantially different from their predecessors.

ITOS, for the first time, offered twenty-four-hour coverage of the earth on a routine, not experimental, basis. Instruments on these craft take cloud-cover pictures at night with a scanning infrared radiometer. The infrared radiation emitted by the earth depends upon the temperature and character of the radiating surface. Rivers and lakes, for example, are cooler than farmland. Clouds stand out vividly at night in infrared pictures, hence, via the radiometer and regular RV cameras, these satellites can "see" cloud patterns day and night.

Infrared, in fact, is becoming one of the weatherman's most valuable tools. Infrared sensing from satellites can pinpoint the temperatures of earth, sea, and clouds to an accuracy within three degrees Fahrenheit. It can accurately estimate how high cloud tops are, thereby detecting their specific type. Infrared readings also can be interpreted to attain a profile of temperatures at various levels in the atmosphere. Such a capability is essential for meteorologists to make more precise longer-range forecasts than were previously possible. Some advanced infrared instruments even measure water vapor at different heights, while others can gauge the total levels of such pollutants as dust, carbon dioxide, nitrous oxide, and sulfur dioxide.

ITOS spacecraft are considerably larger than the earlier Tiros models. They are box-shaped and weigh close to 700 pounds, or more than twice as much as Tiros. An active attitude-control system aboard keeps these satellites' instruments continually pointed at earth. The first operational ITOS was launched December 11, 1970, by NASA, and the second was orbited October 15, 1972. The responsibility for their day-to-day operation is that of the National Oceanic and Atmospheric Administration (NOAA), the successor organization of ESSA.

Covering the globe daily, each ITOS satellite photographs a strip of clouds about 1700 miles wide and 25,000 miles long every two hours, from orbits about 900 miles above the planet. Each picture taken by the TV camera systems covers an approximate 2000-mile square, or about 4 million square miles, of cloud over with a resolution of about two miles. These altitudes, and a spacecraft inclination of 78 degrees, have proved to be close to optimum because the light below for TV pictures is the most consistent at each latitude covered. ITOS pass over roughly the same points at the same time each day. Also, the earth rotates beneath the satellites just enough so that slightly overlapping bands of pictures can be snapped during each revolution.

From the TV pictures scientists can tell much about the position and extent of fronts, cyclones, cyclonic storms, the jet stream, severe weather patterns, tropical storms, sea-ice conditions, and snow cover. In some cases the pictures denote the existence of turbulence, the degree of tropospheric stability, the orientation of surface winds, sea state, and even whether the ground is wet or dry.

Remarkable as the ITOS system is, it still has distinct limitations. While it covers the entire globe daily, it takes twenty-four hours for these satellites to get back to the same spot again. One of the regular features of the earth's weather is the constant changing of its patterns, and often these changes are very rapid. While polar-orbiting satellites (ITOS) play a crucial role in meeting the need for repetitive observation and tracking of large-scale weather patterns, they lack the precision of continuous earth viewing. Many phenomena that directly affect our daily lives exist for such brief periods that polar-orbiting satellites often fail to observe them during the short time they view any one area.

Such information voids will be filled in the 1970s with the advent of the Synchronous Meteorological Satellite (SMS) program. This program is aimed to provide the operational capability to keep significant portions of the earth's cloud cover under *constant* surveillance. These spacecraft will be able to do this because they will be positioned in geostationary orbits 22,300 miles up. From there they can provide full coverage of the earth every twenty to thirty minutes, giving them the capability to furnish the short-term observations required by local forecasters.

These synchronous satellites include an environmental data collection system that will receive information from up to ten thousand sensing platforms located at known remote sites. These can be on land, in ships or buoys at sea, in rivers and lakes, and even in balloons or aircraft. The platforms are designed to relay such information as amounts of rainfall, river and stream heights, wind conditions, air temperature, sea conditions, pressure measurements at local areas, seismic sea conditions, earthquake measurements, and volcanic disturbances.

In other words, SMS spacecraft in orbit will be able to interrogate remote sensing equipment, extract the data from it, and then relay the data to small regional warning and forecast stations all over the world.

The feasibility of operating a weather satellite system from synchronous orbit altitude has been tested and proved by NASA's ubiquitous ATS spacecraft. Four days after its launch on December 7, 1966, ATS-1's spin-scan cloud camera began transmitting essentially continuous photographic coverage of most of the Pacific Basin. Even though this information was being gathered on an experimental basis, it proved to be important in the analysis and forecast activity for that "data-sparse" ocean area for several years. In fact, for meteorologists evaluating weather patterns over the southern hemisphere and the Pacific, pictures from ATS-1 have meant the first significant filling of a virtual information vacuum.

ATS-3 also provided fresh views of weather over much of the North and South Atlantic Ocean area, all of South America, much of North America, and the western edges of Africa and Europe. ATS spacecraft proved too the practicality of collecting data from remotely located ground sensors.

They also have indicated that because they can watch developing weather conditions over a certain area every twenty minutes, instead of just twice a day as in the case of ITOS, synchronous satellites can be an invaluable instrument in storm warnings. ATS cameras can watch a thunderstorm develop from cumulus clouds, and possibly improve the early detection of severe local storms and tornados. Attempts to correlate the photographs with tornado occurrences in the central United States have had some promising results. Pictures from ATS-3 have been received routinely in real time at the National Severe Storms Forecast Center in Kansas City, Missouri. Taken with radar, ground, radiosonde, and other observations, these photographs have contributed to effective tornado warning operations.

Additionally, ATS data has become a regular part of the information available to the National Hurricane Center in Coral Gables, Florida. Used in conjunction with ITOS and other satellite data as well as radar and aircraft reconnaissance reports, the geostationary view can improve the warning operation. The successful tracking of Hurricane Camille by ATS-3 in August 1969 proved to be a definite factor in reliable and timely warnings that were relayed preceding the storm's arrival in the Gulf Coast area.

As soon as it is in operational service, probably in the mid-1970s, the Synchronous Meteorological Satellite program will be run by the National Oceanic and Atmospheric Administration, and will be given a new name: Geostationary Operational Environmental Satellite (GOES). Eventually, a series of perhaps four such spacecraft will be positioned, 90 degrees apart, above the world's major oceans to provide day and night service over the entire planet.

Because weather affects everyone in every country on earth, the United States and more than one hundred other nations, acting within the United Nations, have banded together under the guidance of the World Meteorological Organization and the International Council of Scientific Unions to develop a new understanding of global weather processes and to solve the problem of improved weather forecasting for the benefit of all mankind. This major effort, the Global Atmospheric Research Program (GARP), is explained by NASA's Mathews:

"GARP, as the name implies, applies to all the research efforts being conducted by the world's nations to solve the weather prediction problem. GARP is a scientific endeavor whose central goal is a study of those physical processes that are essential for an understanding of the behavior of the atmosphere which will lead to increasingly accurate weather forecasts over periods from one day to several weeks, and which will lead also to a better understanding of the world's climate."

One of the first major thrusts of this new international program will be the GARP Atlantic Tropical Experiment, to be run through the summer of 1974. Its purpose is to extend man's knowledge and understanding of tropical meteorology, particularly as it affects the weather in the temperate zones. Such knowledge could be used to advance understanding of tropical meteorology, hurricanes, and other storms, to improve forecast techniques for equatorial regions, and to develop the models and observation systems needed to provide reliable numerical forecasts for a week or two in advance.

The experimental area extends approximately from the west coast of Latin America to the east coast of Africa, between latitudes 10 degrees south and 20 degrees north. It includes coastal and island stations in the Caribbean and part of the Gulf of Mexico. In all, twenty-five to thirty ships, a dozen aircraft, and seventy-five land stations, as well as satellites and instrumented ocean buoys, will gather data over the tropical Atlantic and adjacent land areas. It will be the largest and most complex international scientific experiment ever undertaken.

In addition to NASA satellites, instrumented ships from Brazil, Canada, Colombia, France, Germany, Mexico, The Netherlands, Portugal, Russia, the United Kingdom, the United States, and Venezuela will be used in the experiment. Aircraft will be provided by Brazil, Germany, France, Russia, the United Kingdom, and the United States.

Each participating nation will process the data it collects and turn them over to information centers, where they will be catalogued, integrated, and validated. Once the "book" on the GARP Atlantic Tropical Experiment has been completed, the results will be made available to meteorologists in nations all over the world.

The First GARP Global Experiment (FGGE) is planned for 1977. This will involve the world's nations joining together in the quest for improved weather forecasts. The data base will be the most extensive ever collected on a continuing basis. It will employ an improved surface network of stations on land, at sea on buoys, and in the lower atmosphere via balloons. And it will include a spaceborne network of geosynchronous meteorological satellites provided by the United States, Russia, Japan, and the European Space Research Organization. The entire world will be covered, including the North and South poles.

Without satellites, less than 20 percent of the earth's atmosphere can be efficiently observed by conventional meteorological systems.

With satellites, it is feasible to observe and monitor the earth's weather from space for they provide reliable day-and-night observations of the atmosphere on a daily and global basis. With the continuing development of such satellites and their sophisticated sensing systems, and through such exciting international cooperative programs as GARP, experts now believe that before 1980 it will be possible to predict the world's weather reliably up to two weeks in advance.

With such a capability man will at last be able to *do* something about the weather. He will be able to observe and accurately predict the occurrence and locations of floods, tornados, hurricanes, and other types of severe weather, and will have time to issue adequate warnings. He will be able to forecast and to disseminate information on seasonal features and long-term trends. Advance knowledge of an abnormally wet or dry summer, for instance, will enable farmers to plan better for the growing season. Forecasts of an abnormally cold or warm winter would permit proper planning by the power, construction, winter sports, and similar industries. Billions of dollars and thousands of lives will be saved annually.

○
Chapter 7

prospecting from space

We must undertake to increase the performance per pound of the world's resources until they provide all of humanity a high standard of living.

—R. BUCKMINSTER FULLER

The energy crisis . . . the population explosion . . . hunger among hundreds of millions of the world's peoples . . . depletion of natural resources . . . rationing of gasoline . . . brownouts . . . shortages of fresh water . . . pollution alerts.

These are some of the critical problems confronting Spaceship Earth in the 1970s that must be solved if civilization is to progress. Answers have to be found because the problems will compound in the years ahead.

According to the Yearbook of Agriculture, United States population projections range upward from 270 to 380 million by the year 2000. World population, despite international birth control efforts, is expected to soar above 6 billion by the end of this century. To feed the growing masses, experts say, global food production must be doubled by 1985 and tripled by 2000. Even today two-thirds of earth's people are poor, hungry, or undernourished.

Yet in 1955 arable land area per person was about one and a quarter acres, and this is predicted to decrease to little more than half an acre by 2000. Agricultural experts report that as much as 20 percent of world crop production is lost annually to pests and disease. The National Academy of Sciences has estimated that in the United States alone the agricultural losses—to insects, disease, and fire—exceed $13 billion a year. As just one example, in 1970, 15 percent of the American corn crop, some 710 million bushels, were destroyed by corn blight.

It also has been forecast that within the next twenty-five to thirty years, use of water in the United States will increase greatly to support the expanded population. Some scientists have said up to 75 percent of the total average runoff from all the rivers in the country will have to be utilized to meet the increased demands.

During the last thirty years America alone used more minerals and fuels than did the entire world in all of history. And the American consumption of most minerals will double within fifteen to twenty-five years. This country and most other highly industrialized nations already are facing serious shortages of critical mineral raw materials.

All of this emphasizes, with gloomy overtones, the sad fact that earthlings have poorly managed their planet's resources. Only in recent years has man become aware—shockingly aware —that his supplies of air, water, food, and other essentials, long thought to be inexhaustible, are limited. He has learned that the earth's delicate ecological balance can be and is being upset, to the point where all life could perish if corrective measures are not soon taken.

A recent report by the Harvard University Center for Population sums up the stark picture facing future generations: "The 21st Century may witness a world of half-starved, depressed human masses, gasping for air, short of sweet water, struggling to avoid one another, and living at a degraded subsistence level."

"If our ability to find and efficiently utilize resources does not accelerate, and accelerate rapidly, the industrial civilization we now enjoy will crumble within a few decades," warned the late Dr. William T. Pecora, Director of the United States Geological Survey. "The economic status of any nation is almost always a direct function of the use it makes of available natural resources. Put in another way, natural resources are the nonhuman inputs to the economy, and economic growth results in large part from new discovery and effective use of these resources.

"If we had to depend on known supplies, we could predict collapse of our industrial civilization in a couple of decades, and if we had to depend on supplies that could be found and developed, with present knowledge, disintegration would not long be postponed."

To match the acceleration in demand for resources, means must be found to speed acquisition of knowledge concerning them. In short, to survive, to maintain his life style, to progress, man must find a better way to manage Spaceship Earth.

The irony of this is, of course, that there are plentiful life-sustaining supplies on the planet. In the world's oceans, which cover 70 percent of the globe, within the vast arctic tundra and across enormous stretches of desert, in remote, nearly inaccessible mountain ranges, and in the crust and bowels of the earth are great hoards of untapped resources. There are mineral riches, energy sources, and food and water stores that far exceed the present and future needs of the world's billions of people. But these resources can support the ever-accelerating demands only if imaginative new ways are found to discover them effectively. Man must develop new means, new tools to locate them.

The Space Age has provided such a means. The tools are man-made satellites.

Laden with cameras capable of ultrahigh resolution and with sensitive measuring instrumentation, spacecraft in orbit several hundred miles up can take the earth's pulse electronically. They see what man cannot. Not only can they survey and inventory the world's resources more economically and effectively than ever before, but they can also detect elusive clues that will lead man to new supply sources. The potential applications of such satellites are virtually limitless.

AGRICULTURE

In the field of agriculture, satellites could photograph every sizable farm in the world to determine what crop was being raised and whether the crop is young or old, healthy or diseased, and accurately predict the yield. They could also send advance warnings of droughts or even of changes in soil condition to aid in the prevention of blight, and similarly, they can detect the larvae of locusts. Such capabilities will allow farmers to decrease their use of polluting chemicals by "telling" them when and where to spray most effectively. So great is the loss in the United States alone because of plant diseases and insects that it has been estimated that just a one-percent reduction in crop loss could amount to as much as $75 million in savings annually. Other agricultural uses could include analysis of soil moisture content, determination of irrigation requirements, and timely censuses of livestock.

Former Secretary of Agriculture Orville Freeman has predicted what the farmer of the future can expect: "Information gathered from throughout the world (via satellite) is transmitted to computers for analysis and immediate use. The soils of the world have been inventoried, and each crop is grown either on the soil best suited for it, or on soil chemically modified for maximum productivity. We have a running inventory of acreage and output of all crops, and we use accurate predictions to guide marketing and distribution to avoid waste and local shortages and surpluses."

FORESTRY

Like the mountains and seas, the earth's billions of acres of woodlands are a storehouse of raw materials so vast that it affects the weather. Birds, beasts, and humans have long depended on forests for shelter, and wood product industries are a major segment of the international economy.

The Forest Service of the United States Department of Agriculture keeps a continuing inventory of the country's timber to provide volume, growth, and drain statistics to local and national planners and legislators. Some data are as much as nine years old when reported, however, because of the difficulties and cost of collecting information about woodlands.

To foresters, a satellite system is the only practical way to mount a constant watch over vast wooded areas to provide warning of insect infestations and diseased trees, take censuses of trees, and report logging yields. Instruments in space can spot forest fires burning in places man has never seen, and they will be able to pinpoint the boundaries of active fires throughout the thickest smoke and haze, day and night, in any weather.

Such data will provide direct economic benefits. It costs more than $150 million a year to control forest fires in the United States. They cause tangible property damage of approximately $500 million annually, and countless deaths and injuries. And they cause destruction of soil, water, wildlife, and other resources whose values are less tangible.

Surprisingly, insects and disease account for timber losses equal to annual growth and exceeding by 700 percent the losses due to fire. To determine the location and extent of major outbreaks, the Forest Service spends more than $5 million annually on insect and disease surveys. Satellites will not only be able to do this job far better, but also at substantially less cost.

LAND MANAGEMENT

Continuous observations from space will prove a great boon to techniques of land management. National parks and forests and public lands comprise 175 million acres in the continental United States and nearly 290 million acres in Alaska. They are a rich resource base for the future, in addition to yielding current dividends to the nation in oil, gas, forest products, and recreation. These lands also support about 7 million head of livestock and nearly 3 million big-game animals. Conservation of these irreplaceable resources helps support the $20-billion-a-year outdoor recreation industry.

The data-gathering potential of orbiting satellites can help the Department of the Interior in administering these public lands and preserving their ecology in a number of ways. Studies of changing features or conditions such as grassland status and foraging patterns could be supported by synoptic observations from space. Environmental management can benefit from timely and reliable information collected and transmitted from space on the distribution, health, and vigor of vegetation and the measurements of snow accumulation and glacier movement.

CARTOGRAPHY

Cartographers admit that 70 percent of the world's small-scale maps are considered inadequate, and the remaining 30 percent are obsolete. In the United States, where aerial photography is done more systematically than in most other areas, some parts of the country have not been photographed in twenty years, and even the most modern national land maps are reportedly at least ten years out of date.

To make an accurate photomosaic of the entire United States, using airplanes, approximately a million aerial photographs would be required, and it would cost about $60 million to rectify them for geometric distortion. From satellite altitudes such a photomosaic would require only four hundred pictures and could be assembled in a few weeks.

Detailed maps of rural and urban areas, kept up-to-date, will have many applications in helping planners develop new traffic arteries and terminals. Surveys from space of urban areas, indicating housing and population densities, park areas, industrial development, and types of settlement will prove invaluable to renewal and new building programs.

GEOLOGY

Some of the most dramatic and immediately rewarding uses of earth information obtained from observation platforms in orbit will be in geology. The Department of the Interior is keenly interested, for example, in the exciting possibilities of discovering new mineral and energy sources. Most metallic mineral deposits and subterranean gas and oil stores are associated with such earth structures as rock folds and faults. Prospecting from space will permit charting of known resource areas and discovery of new ones in remote areas. While it is impossible for them actually to detect minerals, space satellites can help locate areas that look promising because of their large structural features. Geologists can then investigate those areas further from the ground or through aerial photography.

Aerial photographs have already proved useful in revealing telltale signs that lead to virgin mineral riches. In Canada hundreds of thousands of miles of land have been surveyed from camera-carrying aircraft, and the photographs have yielded such spectacular finds as the Manitoba nickel deposits of International Nickel and the base-metal discovery of Texas Gulf Sulphur in the Timmins, Ontario, region. Aeromagnetic surveys of basement rocks in the Pea Ridge area of Missouri led to the discovery of an iron-ore deposit 1300 feet beneath the surface that has been valued at $2 billion.

Satellites will be able to do jobs like these better and faster, everywhere in the world rather than over relatively small strips of land. Already pictures from orbital altitudes have proved far superior for viewing the planet's larger linear geological features.

Such observations from space may be useful to scientists in other ways. Oxidation of some mineral deposits generates heat; consequently, the deposits can be detected by infrared sensors, and repetitive satellite coverage of the melting patterns of snow may also point to new lodes.

Through the use of ultraviolet techniques, spaceborne instrumentation may aid in the finding of phosphate "glows" in remote areas of the world, betraying the location of such deposits and helping solve the need for fertilizer in nations where food supplies are short.

Another geological application will be in the critical search for new power sources. Infrared instruments on aircraft have detected temperature variations at or near the earth's surface which may indicate potential sources of geothermal power.

Continued scanning of the earth by satellite may disclose crust movements and thermal gradients which could provide timely data for predicting such natural disasters as earthquakes, volcanic eruptions, and landslides.

In volcanoes of the Hawaiian type, for example, the earth actually swells prior to an eruption as lava and gas are forced upward. The swelling is measured with sensitive levels—one of the methods Geological Survey experts use to predict an eruption. Scientists believe that infrared sensing from space of heat patterns on volcanoes will greatly aid in their forecasting of major volcano activity.

. In addition to sensing devices, long-range satellite telemetry would make a significant contribution to programs aimed at earthquake prediction, which would save hundreds of thousands of lives and billions of dollars.

OCEANOGRAPHY

The world's oceans are man's greatest and least-used natural resource. Covering over 133 million square miles of the planet, this very vastness has made it impractical to undertake continuous broad-scale surveillance of them by conventional means. The job can be done effectively only by satellite.

Not only do the oceans serve as a prime source of food and minerals, but they also directly affect earth's weather. It becomes essential, therefore, to develop a means to monitor such oceanographic features as sea state, distribution of sea ice, surface temperatures, current patterns, and biological developments. This will also provide information vital to the shipping trade.

Infrared instruments from orbital platforms can be used to trace the temperature outlines of ocean currents and upwellings, and can help lead man to one of his major food supplies—fish. Surface temperatures can help identify the highest plankton concentrations—the prime source of food for fish. In early experiments satellites have proved that this can be done by leading Taiwan fishermen to an area previously undiscovered, but teeming with edible marine life.

HYDROLOGY

In water resources satellites can perform a number of essential services. The world's oceans, ice fields, lakes, and rivers contain more than 324 million cubic miles of water. Beneath the earth, in soil and rock, lie some 2 million cubic miles in the form of ground water, much of it saline. Another 3100 cubic miles of water, mostly as vapor, is contained in the earth's atmosphere.

Scientists say most of the earth's original supply of water is probably still in use: little has been added or lost in the hundreds of millions of years since the first cloud formed and the first rains fell. The same water has been pumped time and again from the oceans into the atmosphere, dropped upon the land, and returned to sea. In this hydrologic cycle, at any instant, only about 5 of every 100,000 gallons of the total water supply is in motion: most of the water is stored in the oceans or underground, frozen in glaciers, or held in lakes. Television photography and instrument measurements made from space will help man gain a better understanding of the world's "water bank."

Satellites also can be used to:

- inventory water in regional basins by measurement of lake levels, river flow rates, irrigation patterns, and drainage patterns;
- provide early warning of floods by monitoring rainfall and surveying drainage basins;
- locate aquifers and determine the suitability of various sites for constructing dams and storing water;
- estimate water resources through snow and frozen-water surveys, and determine the location of seepage and other ground water sources. It is known that vast quantities of ground water escape from continents and islands into the oceans. The ability to view island margins from space will enable hydrologists to pinpoint areas of large concentrated ground water "leakage," and cut down on waste.

There are other, related applications for an operational network of observational satellites in earth orbit. Cameras with high-resolution capability might reveal cultural features of interest in desert areas or even promising archeological sites to explore.

Spacecraft may have wide-ranging uses in the control of some of man's oldest and most dread physical afflictions. The correlation of such malaria-producing phenomena as land/water distribution and land and water temperature could be applied for quick and accurate combating of this disease.

One of the most intriguing possible missions for orbiting electronic eyes may be in the detection of such illegal crops as opium poppies. Since sensitive instruments can determine the type, age, and health of crops, it is reasonable to assume that they could also locate marijuana and poppy fields.

NASA and other government agencies have long recognized the great potential of satellite observations of earth in all these areas. Photography experiments from the manned Gemini and early Apollo flights in the mid- and late 1960s demonstrated the importance of surveying the world's resources from space. Consequently, when NASA solicited proposals for experiments to be conducted on the first satellite in the program, the response from scientific, industrial, and governmental communities worldwide was overwhelming.

More than three hundred experimenters, representing the United States, thirty-seven other countries, and two United Nations groups, were selected for participation in the first flight. The breakout included eighty-three principal investigators working on mineral resources, geological structure, and land-form surveys; sixty-two on agriculture/forestry range resources; forty-five on water resources; forty on land-use surveys and mapping; thirty-seven on environmental studies; and twenty-nine on marine and ocean surveys.

Other experimenters were interested in the development and potential of the cameras and instruments to be used.

Thus it was with great worldwide scientific interest that ERTS-1 was launched July 23, 1972, and successfully placed in orbit 575 miles above earth. The spacecraft, built by the Space Division of the General Electric Company, looks much like a Nimbus meteorological satellite, on whose physical design technology it actually drew heavily. Some telemetry, tracking, and command equipment design was derived from Apollo spacecraft systems. This philosophy of using, to the maximum practical extent, designs and hardware previously developed and used in space programs was instituted to produce an economical, reliable spacecraft whose schedule from drawing board to first flight was a tight two years.

From its all-seeing vantage point ERTS-1 circles the earth every 103 minutes, or 14 times each day, covering everything except small areas around the North and South Poles. Because the orbital plane of ERTS-1 is not in exact synchronization with the earth's 24-hour rotation period, there is a 25.8-degree "westward shift" on each of the satellite's successive orbits. In other

Dumbo-eared Earth Resources Technology Satellite—ERTS-1—covers most of the earth's surface with its sensitive instrumentation, which relays to ground stations important data relating to agriculture, forestry, land use, geology, hydrology, geography, oceanography, and environmental science.

words, on one day it will pass over Maine during a morning pass, over central Minnesota on its next swing around, and over the western portion of Washington state on the next, and so on.

It views 115-mile-long north-south swaths of earth as it orbits. The flight path is such that the satellite completes a full cycle of coverage every 18 days and then begins a repeat. This provides progressive information on specific sites—the ripening of harvests can be followed, for example.

ERTS-1 carries television cameras as well as radiometric scanners to obtain image data on various spectral ranges of visible light (red, blue, green) and infrared. These can record a wide range of phenomena invisible to human eyes, undetectable by human senses. The three Return Beam Vidicon Television cameras produce photos of the planet's surface in squares about 115 statute miles square, with a resolution of about 300 feet. Each camera views the same ground scene, but in a different spectral band, in the visible and near infrared portions of the spectrum.

Since this depends on reflected solar energy, the cameras can be operated only during the daylight portion of the orbit. All three cameras are mounted on a common baseplate for mechanical stability and then shutters are synchronized; the earth-surface image is stored on the photosensitive face plates of vidicon tubes.

The Multispectral Scanner System gathers data by simul-

Rich, rugged features of the Grand Teton National Park area and Idaho Falls, Idaho, were captured in this brilliantly clear composite photo taken from NASA's ERTS-1 satellite.

taneously imaging the surface of the earth in four spectral bands through the same optical system. In ground processing the Scanner frames—also covering squares of 115 miles square—are constructed from the continuous strip to correspond with the images produced by the television camera system.

On ERTS-1, electronic instrumentation is commonly referred to as remote-sensing systems. Literally translated, that means it has the capability of detecting the nature of an object on earth without actually touching it. From space sensors probe or "listen to" an object electronically, then convert the electronic signals to a visual record—a photolike image.

Remote sensing is made possible by the simple physical fact that any object whose temperature is above absolute zero will reflect, emit, transmit, absorb, or scatter protons—the basic units of electromagnetic energy.

Across the visible and invisible spectra all objects yield distinctive "fingerprints," or spectral signatures, that are determined by the objects' atomic and molecular structures: wheat, for example, has a different signature from corn or oats. Moreover, identification of these signatures enables scientists to determine not only what an object is, but how old and how healthy as well. The cell of a sick plant reflects or emits radiation differently from a healthy cell.

"False color" techniques are used to enhance the images, helping investigators to analyze the nature of the subject matter photographed in great detail. Green, red, and infrared, seen and recorded separately by the satellite, are combined at NASA's Goddard Space Flight Center in Greenbelt, Maryland. Thus, healthy crops, trees, and other green plants, which are very bright in infrared but invisible to the naked eye, appear as bright red. Suburban areas with sparse vegetation appear as light pink, barren lands as light gray. Cities and industrial areas show as green or dark gray, and clear water appears black.

When the green, red, and infrared are combined in different ways, the image can be analyzed to provide information about the quality, condition, and kinds of objects in the scene in more detail than could be obtained from conventional photography. In fact, one of the most promising uses of the technique is the development of automated data-processing systems that will permit the direct translation of the satellite-gathered data into maps or charts defining land-use categories and the areas devoted to each—such as crop types and acreages, forest areas, urban boundaries, and so on.

As a sort of "bonus" service, data-collection systems aboard the spacecraft gather information from unmanned collection platforms scattered at remote sites throughout the United States and its coastal regions. These data may include periodic sampling of such local environmental or surface conditions as temperature, humidity, stream flow, and soil moisture.

When the spacecraft is in view of a transmitting platform and a ground receiving station, the message is relayed immediately to the station through the satellite. Otherwise, the message is stored by onboard tape recorders to be transmitted later. These data are subsequently passed on to individual investigators for detailed analysis.

A data subsystem links the observatory in space with the users on the ground. Information collected by the satellite sensors is converted to electronic signals, processed, and transmitted to earth stations. Here it is received, classified, processed, stored, and/or provided immediately to investigators and other users all over the world.

The first Earth Resources Technology Satellite has been successful beyond even the most optimistic expectations. "The ease with which ERTS-1 is obtaining accurate, detailed, and useful information about large regions of the earth every eighteen days is unprecedented," says Charles W. Mathews. "During the first four months of [ERTS-1's] service, the North American continent was covered many times, and large portions of all other continents, coastlines, and polar regions were observed at least once, with repeat coverage continuing.

"The quality of the photographlike images is excellent. Early results of data interpretation indicate that a unique and highly successful demonstration of the ability to meet the desired information needs is under way. The large area covered by the images is a unique and valuable aspect of the ERTS-1 data, since it allows the recognition of surface features which are too subtle to be found on even the best mosaic pictures. Coverage and detail exceed our expectation. Linear features such as roads, pipelines, canals, etc., can easily be seen."

An average of about 4 million square miles of earth are mapped daily by the satellite's sensor system. During the first six months of operation ERTS-1 imaged nearly 40,000 scenes of the earth's surface. From these about 1.5 million high-quality photographic images have been made and shipped, from the central data-processing facility at NASA's Goddard Center to government agencies and investigators all over the world.

(Incidentally, the Department of the Interior maintains a national repository and dissemination agency for data from ERTS satellites and other remote sensing equipment at a new center in Sioux Falls, South Dakota. It is managed by the United States Geological Survey. Any government agency, private business, or individual citizen can purchase, at nominal cost, satellite pictures of virtually any specific site on earth. The many users of these photographs range from land development firms that need them for studies to people who enjoy having satellite photographs of their hometowns as unique souvenirs.)

The examination of the data is still going on, amid much excitement and enthusiasm over the early results. In its first few months in space, for example, the satellite exposed timbers in Oklahoma that had been damaged by chemicals, revealed undiscovered geological faults in California's Monterey region, and led geologists to traces of an old volcano in Nevada that subterranean forces seemed to be lifting.

The spacecraft's discoveries were not limited to the United States. After examining the first imagery of his country, Dr. Fernando de Mendonca, general director of Brazil's National Space Institute, said, "The ERTS-1 imagery over the Amazon region has indicated to us some extraordinary conclusions." In a report he summed some of them up: "The courses of the tributaries of the Amazon River are very different from the ones shown in the most recent available charts. Islands of more than 200 kilometers square exist which are not shown on maps. Small villages and towns are located incorrectly on the maps by several tens of kilometers. The drainage systems of some areas are entirely wrong, and this has caused, among other things, the construction of roads with extra expenditure for bridges.

"The entire Amazonian region was covered last year with Side-Looking Airborne Radar (SLAR)," Mendonca continued. The completed controlled photomosaics will not be ready for at least another year. Over 150 people are working on the SLAR project, which has cost Brazil about $20 million. The cost of ERTS imagery per square kilometer is about two orders of magnitude less than the SLAR if the satellite operates for the expected lifetime of one year." Actually it has far exceeded that life expectancy.

In March 1973 more than one thousand scientists and experts, including most principal investigators, met in New Carrollton, Maryland, at an ERTS symposium to discuss some of their findings from imagery gathered during the first few months of the satellite's operation. More than two hundred technical papers were presented at the symposium. The following is a cross-section sample of one of the more interesting findings:

• Marian Baumgardner, of the Purdue University Laboratory for the Applications of Remote Sensing, reported on an ERTS study over Lynn County in northwestern Texas aimed at identifying, characterizing, and mapping soils, vegetation, and water resources in a semiarid region. This is of paramount importance because more than a third of the world's lands lie in such regions.

Baumgardner said the data suggested that croplands could be mapped and measured, as could vegetative, species, and management differences in rangeland. Gross soils patterns and differences, related to agricultural and land-use management problems and practices, could be ascertained from space and mapped. From satellite-survey information crops damaged or destroyed by hail and windstorms could be mapped and measured, as could areas of bare soil with the related problems of erosion and the need for conservation.

Of what value is such information? Baumgardner answered this question in his paper: "Information relating to areas of crops planted and harvested is valuable to various agencies of the United States Department of Agriculture, [such as] the Federal

Crop Insurance Agency. [Also] for agricultural industries, especially to the chemical ones.

"What about areas of crops damaged and destroyed by hailstorms and windstorms? The Federal Crop Insurance Agency, private insurance agencies, and in this particular study the High Plains Cotton Council." (Cotton accounts for approximately 58 percent of the area's crops, and the principal income of Lynn County is derived from farming and ranching.)

"With regard to soils patterns and delineation of soils differences, this information is valuable to the Soil Conservation Service, to land-use planning agencies, to county, regional, and state planning groups, and to state agencies concerned with natural resources. Surface water information—surface maps delineating seasonal changes—is of great importance to the High Plains water district, to the Texas Water Development Board, to the U.S. Geological Survey, and to any other agency concerned with water resources management."

The Baumgardner report concluded: "These early results from the computer-implemented analysis of ERTS data suggest several valuable tasks which may be performed in the semiarid [and] arid regions of the world: assessment of conditions of croplands and rangelands, mapping and measuring crops and ranges, monitoring of water resources, and rapid delineation of soils boundaries."

Possible new sources of oil deposits were disclosed in a paper presented by the Geological Survey's William Fischer, who discussed ERTS imagery over an area in northern Alaska not far from Point Barrow. "We have here a coincidence of gravity data, magnetic data, and ERTS observations that would lead us to believe that we are seeing the reflections of very deep-seated structures in the overlying, very young rocks. If so, this then becomes a major and heretofore unobtainable bit of information that can aid . . . the development of the oil provinces of the north.

"Alaska's land mass covers 586,000 square miles. . . . The basic data for informed land-use research and planning in Alaska is [sic] sparse and . . . often outdated. . . . Alaska is so vast and the arctic environment is so varied that this environmental knowledge gap will not be bridged very quickly by conventional means or with normal dollar resources.

"This is why the ERTS program, [with] its demonstrated capability for economical large-scale surveys, affords a unique opportunity to narrow this knowledge gap. . . . I think it will accelerate our exploration for mineral deposits and for petroleum."

The Fischer report pointed out other interesting satellite applications: "One area of environmental concern is the northeast Alaska caribou population. These animals number approximately 150,000 at present, but there have been large fluctuations in herd sizes over the past fifty years. Migration routes and winter dispersal patterns are not yet well enough known to significantly improve the management of the caribou resource at the present time. Snow cover has long been recognized as a major factor influencing the biology of caribou, but aerial surveys to obtain these data over the vast areas are prohibitively costly.

"ERTS will help us to get a handle on this problem for the very first time," Fischer said. ERTS images are being used in two ways to monitor herd activity. One is to map habitat favorable to caribou, and the other is to locate and map environmental features that arise from large caribou aggregations."

NASA previously has demonstrated successfully the satellite tracking of animals equipped with miniature radio packs. In one experiment scientists were able to locate and map the movements of an instrumented elk during a twenty-eight-day period. Similar experiments are not being conducted with bears; instruments monitor them when they are hibernating or moving.

Another fascinating satellite study of Alaska involves snow, which "has many adverse effects on man's activities in the arctic and the subarctic because it remains on the ground for long periods of time," says Fischer. "But snow is also beneficial in that it thermally insulates the soil-atmosphere interface and affords protection to plants and animals. And it is a means of storage of great quantities of water, which then runs off during the summer.

"So, from ERTS imagery, we can for the first time produce maps of snow lines across Alaska during the initiation and the decay of the seasonal snow cover."

The Fischer paper told also of an ERTS study of a wide-ranging spruce beetle infestation in a region near the Cook Inlet Tyonek Indian Reservation and in the Kenai Peninsula. It is estimated that perhaps 200,000 acres are involved and as much as 2 million board-feet of lumber. Timely data on the extent of the damage and up-to-date inventories on healthy stands and new kill and old kill areas are essential to agencies charged with the management of forest inventories. But to maintain surveillance of such a vast rugged area would be nearly impossible without the sensitive sweeping eyes of satellites.

"The applications of ERTS data," Fischer summed up, "are playing extremely vital and timely roles in the planning for the imminent, and, we hope, orderly development of Alaska."

The United States Geological Survey has conducted a major investigation into the usefulness of ERTS data for estimating damage and for mapping the extent of the disastrous 1973 flood of the Mississippi River. Satellite images clearly showed the extent of the actual flooding, while areas that had been flooded and subsequently drained could be delineated with reasonable certainty. The Geological Survey's special report observed that "repetitive coverage is necessary for analysis of a disaster of this type."

The papers presented at the symposium outlined hundreds of other uses of ERTS data for the direct and immediate benefit of man on earth, among them were:

- Imagery pinpointing concentrations of menhaden fish in the Mississippi Sound. These data are available to fishermen on a daily basis as they leave for their day's fishing. Since it is estimated that about 70 percent of the cost of a fishing operation is getting to where the fish are, their location is of great importance to fishermen.
- The ERTS image of a part of the Gulf of Mexico coastline of Mississippi and Louisiana revealed a number of arcuate and circular features ranging from six to forty miles in diameter. Comparison of these features with existing tectonic maps of the region indicates that in the inland areas, shown on the image, the location of many of the features corresponds to known salt domes. There are, however, similar-appearing features in the near-coastal area that are not shown on the existing tectonic maps, suggesting that additional exploration for salt domes in the coastal zone may be worthwhile.

• In southern Arizona ERTS was used to study accelerated erosion, which since 1880 has been devastating extremely valuable grazing land. Successive ERTS plotting is making possible for the first time a study of the underlying causes of such erosion.

• Satellite-relayed information in 1973 helped the Salt River Project in Arizona through the wettest runoff season in thirty-two years. Watershed specialists there said ERTS provided as much as twenty-four to thirty-six hours' advance notice of heavy inflow to the reservoir. This gave water managers time to make adjustments that minimized necessary release of water into the Salt River Channel.

• ERTS-1 data are permitting rapid and more complete assessment of such physiographic parameters as drainage areas, stream-network character, vegetation cover, and surface water features in large and remote regions in west-central South America, Spain, Vietnam's lower Mekong River Basin, and the Republic of Mali. A specific example: a 170,000-square-mile region in South America hundreds of drainage basins were inventoried, using thirty-one images from ERTS with reference to surface water features. Thirty-six new lakes were discovered. The satellite also detected that in the lower Mekong River Basin 4 million hectares of primary forests were seriously degraded by war action, obviously altering the region's natural hydrological cycle.

• In southern Florida water level and rainfall information is collected by ERTS and relayed to water-resources management agencies in thirty to forty-five minutes! This has enormous impact, for it permits experts to make accurate on-the-spot assessments of the state of the delicate ecological balance in an area involving such vanishing species as alligators; an area critically dependent upon existing water supplies.

• In Mississippi and other states ERTS high-resolution imagery distinguishes entire transportation networks—highways, interstate systems, primary and secondary roads, railroads, power lines and pipelines—and the data are being applied to land-use management and urban planning, from deciding on school bus routes to adding to the tax rolls homes that were not known to the tax assessor.

New applications for information gathered by the satellite are found almost daily. "The African nation of Mali plans to develop its first definitive water-resources inventory using ERTS data," says Dr. Fletcher. Another hydrological use of ERTS is the measurement of snow cover. The satellite can photograph the Grand Tetons area every eighteen days, and as the season progresses, the increase in snow cover can be measured and plans made for handling the spring runoff.

One of the most exciting aspects of the ERTS program is the "unknown" factor: discoveries about earth made by satellites in space. An example of this occurred in 1973, when the Geological Survey reported the finding, from sensor imagery, of two sets of mountains, hitherto unknown, unmapped, and unexplored, in the Lambert Glacier area of eastern Antarctica and west of the Prince Albert Mountains in southern Victoria Land. It observed that such space imagery "shows enormous potential" for polar-region mapping and for a variety of earth science studies in Antarctica.

Imagery from cameras aboard ERTS-1 and Skylab also pro-

Splendid portrait of Greece was made by all-seeing "eyes" of NASA's Earth Resources Technology Satellite from an altitude of 569 miles. Many significant geological features, such as folds and faults, are clearly visible.

vides scientists with the tools necessary to locate and map fault areas and geological structures normally associated with earthquakes. In one such study scientists at Rockwell International's Science Center in Thousand Oaks, California, are examining photographs covering an area running east from San Francisco to Denver, southeast to the Gulf of Mexico, and west through Sonora, Mexico, and Baja California to the Pacific. This technique can be used to delineate areas susceptible to earthquake recurrence which seismic data may misleadingly indicate to be safe.

Working under a NASA contract, experts use blowups of specific areas photographed from space, then chart on them all known faults and major geological features. They carefully examine the imagery for "hidden" geological structures or other clues not previously known and therefore not mapped. This has special significance in such remote, sparsely populated, seldom traveled areas as the Nevada and Arizona deserts and wastelands. Dr. Mohammed Abdel-Gawad, Rockwell's principal investigator, reports that an astonishing 100 percent more faults than were previously known have been discovered during the investigations.

Such revolutionary findings have become almost routine for ERTS since its launching in July 1972, and satellite imaging systems have exhibited enormous potential. While ERTS-1 is experimental, it has proved beyond all expectations its value as a tool with which man can better manage his planet, make more efficient use of its munificent resources, and improve his quality of life.

In July 1973 the satellite successfully completed its first full year of operation, having provided more than 70,000 photographs of three-quarters of the earth's surface and virtually all of the United States land mass. NASA plans to launch a second experimental Earth Resources Technology Satellite in 1976, possibly earlier, which, in addition to collecting data to extend the ERTS-1 type of investigation and studying long-term trends of the consequences of human interaction with the environment, will carry advanced instruments to increase its capabilities.

Following the second experimental flight, an operational satellite system will be developed. Editorializing in *Aviation Week and Space Technology*, editor Robert Hotz wrote, "The

ERTS experiment has shown—at a relatively modest cost of $160 million—that it is feasible to measure, catalogue, and monitor the contents of the earth's crust. ERTS is, in fact, the only way that a comprehensive index can be compiled of the earth's natural resources along with the forces that are eroding and consuming them. It has also yielded more pertinent information faster and cheaper than all previous methods of tackling this job in the limited areas where it was possible.

"ERTS has pointed the way for man to do a far more effective job of managing the finite resources of his tiny unique planet. There should be no delay, either by the U.S. government or the rest of the world, in making the financial and technical contributions required to provide a complete operational ERTS system as soon as possible for the service of all mankind."

○
Chapter 8
pollution detectives

The vast loneliness up here is awe-inspiring, and it makes you realize just what you have back there on earth. The earth from here is a grand oasis in the big vastness of space.

—ASTRONAUT JAMES LOVELL
IN ORBIT AROUND THE MOON

Under the cloak of darkness a giant freighter dumps tons of raw sewage into a harbor area near the shoreline. A few miles farther out a tanker blows out its bilges, emptying hundreds of tons of filmy oil-slick water into the sea. Preparing them for the scrap-heap, junk dealers in large cities illegally burn old cars at night to hide the choking curls of toxic smoke. Across the nation large factories store their liquid effluents produced during daytime operations and discharge them into nearby bodies of water after dark, clogging the streams and destroying marine life. Power-generating stations blatantly pour torrents of steaming water into the sea.

These are but a few examples of how the world's water and airways are daily being fouled by irresponsible polluters who defy the law and care nothing about the long-term consequences of their actions.

As a result America's creeks, rivers, lakes, seas, gulfs, and oceans are becoming contaminated beyond redemption through increasing force-fed diets of radioactive waste, gases and irritants, heavy metals, trace elements, and chemical affluents in innumerable forms including lead, mercury, pesticides, oil solid wastes, raw sewage, cleaning fluids, detergents, and all the poisonous byproduces of overpopulated areas.

This triggers a chain reaction with catastrophic potential. Oil spills can cause massive destruction of marine life and deep sea organisms. Lake Erie, for example, has been glutted each year with 2.5 million tons of silt, sewage, and such industrial wastes as pickling acids from the steel mills and phosphate-based detergents. The biochemical oxygen demand of this overload has exhausted the supply of dissolved oxygen and the lake is now biologically dead. The pollution situation was so bad in Ohio's oily Cuyahoga River that when it actually caught fire in 1969, the blaze raged out of control for days and burned two railroad bridges.

Air pollution consists of toxic gases introduced into the atmosphere: carbon dioxide, particles like fly ash, volcanic or radioactive dust, and aerosols used to disseminate pesticides. Distribution is worldwide. Lung-searing combustion products from industrial processes and the operation of aircraft and automotive engines introduce carbon monoxide, hydrocarbons, lead compounds, sulfur dioxide, and nitrogen oxides. Sufficient quantities of these gases can alter the chemical composition of localized atmospheres.

Scientists estimate that the atmosphere contains about 500 million tons of carbon monoxide. Each year automobile exhaust, industrial activities, and other sources generate some 200 million tons of this gas. The air has become so bad in some cities that in Tokyo, for instance, traffic policemen have to take "oxygen breaks" and inhale oxygen from tanks to catch their breath.

"We now have 50 percent more nitrogen oxides in the air in California [than 25 years ago]," says ecologist Kenneth E. M. F. Watt. "This has a direct bearing on the quality of light hitting the surface of the earth. At the present rate of nitrogen buildup, it's only a matter of time before light will be filtered out of the atmosphere and none of our land will be usable."

The overall effects of this mounting pollution, if not checked, scientists warn, could disrupt the oxygen and carbon dioxide cycles vital to human life, and threaten the entire ecological balance of the planet. Some ecologists are convinced that man's pollution is building up a layer of particles in the atmosphere that, together with volcanic dust, blocks more and more of the sun's energy, and thus presents the grim possibility of another Ice Age. Airborne particle pollution has doubled in the northern hemisphere since 1910, because of dust, smoke, and the invisible particles in automobile exhaust. And the rate of such pollution is rapidly increasing.

The monumental problem of stemming the pollution tide and protecting the precious environment is not limited to any one nation, continent, or hemisphere. It is global. George F. Kennan observed that ". . . the national perspective is not the only one from which this problem [of preventing a world wasteland] needs to be approached. Polluted air does not hang forever over the country in which the pollution occurs. The contamination of coastal waters does not long remain solely the problem of the nation in whose waters it has its origin. Wildlife—fish, fowl, and animal—is no respecter of national boundaries, either in its movements or in the sources from which it draws its being. Indeed, the entire ecology of the planet is not arranged in national compartments; and whoever interferes seriously with it anywhere is doing something that is almost invariably a serious concern to the international community at large."

The problems of worldwide pollution and its harmful effects have become fairly well recognized in the past few years. The big question is, What can be done? It is obvious that there is a very real need for a global pollution watch, for a system that not only can detect and report flagrant violators, but also can monitor the planet's ecology.

The answer, again, may be in space.

As early as 1966 the United States Department of the Interior, recognizing the potential of unmanned satellites in earth orbit for environmental purposes, issued a news release that said, in part: "There are four major facets to water pollution problems: (1) detection of the presence of a pollutant; (2)identification of the specific pollutant; (3) measurement of the pollutant's concentration; and (4) determining the movement and fate of the pollutant in the water environment.

"Orbital sensing of large bodies of water such as Lake Erie or the Caspian Sea will aid immeasurably in mapping the water-flow patterns, the distribution, and fate of polluting substances."

"One of the real advantages of a spacecraft system is its capability for fast wide-scale information-gathering," says Dr. Carl D. Graves, an earth-resources project manager with the TRW Company. "Minnesota, for example, is 'the Land of 10,000 Lakes.' To monitor these lakes from the ground would be . . . very costly and time-consuming. Spacecraft could do this kind of job easily, giving us the information we need to keep Minnesota's lakes clean and usable for sport and recreation."

Orbiting satellites will be able to cover the entire United States continuously and automatically, detecting and identifying different pollutants as they move into water. Violations can be pinpointed as they occur, on major rivers and lakes down to fifty acres in area. Ground crews can then take samples and identify the causes of the pollution.

Satellites will also be able to track air pollution and its distribution patterns over great distances, identifying concentration levels and rates of movement and dispersion. It is not enough to monitor the air over individual cities—Los Angeles smog, for example, sometimes appears over Arizona. Regional information is needed to answer such questions as: How does Chicago's pollution affect Detroit, or Pittsburgh, or St. Louis? Where does the dirty air originate that appears in Montana and other Western states? How far out into the Atlantic does the pollution from the northeast corridor area reach? With their ability to give frequent pictures of large areas, spacecraft can provide such information on an international basis.

Orbiting satellites can also serve as data-relay stations. Receiving information from small electronic monitoring devices implanted in lakes, rivers, and coastal ocean areas, ground sensors would transmit to satellites information on water content and pollution trends which would then be relayed to a data-processing center on earth for evaluation. Planners on the ground would be able to use this information to deal with the pollution. "The real point is not that a spacecraft at 500 miles' altitude can stop pollution," says Dr. Graves, "but that it can give us, quickly and frequently, information on the nature of that pollution." While it is theoretically possible to monitor the oceans or the Great Lakes from the ground, the vast resources needed to do the job—ships, manpower, information networks, and other elements—render the task impractical and prohibitively expensive. "Here is a case where spacecraft are clearly a much more cost-effective way to do the job," says Dr. Graves. "In short, a spacecraft system can provide information on a large scale, on a frequent and repetitive basis—even in remote areas."

How would spaceborne systems operate? Much as the cameras and instrumentation on the ERTS-1 satellite. To keep effective track of water pollution, a variety of hydrological characteristics must be measured: surface-temperature gradients in lakes and streams, sedimentation dynamics, precipitation, lake and reservoir levels, and tonal colors.

Differences in water color may correlate with chemistry and such vegetation as plankton bloom and algae, thereby contributing to pollution studies. And polluted water, which is likely to be warmer than adjacent clean water, could be detected by infrared scanners that produce images similar to photographs that distinguish water temperature differences to a tenth of a degree utilizing gray-scale tone—the darker the gray, the colder the water. Varying water temperatures can thus be determined, allowing a better understanding of the rate at which pollutants move and of their distribution and dispersion within the water.

The multispectral device photographs the visible electromagnetic spectrum. Certain discharges, metals or particulates, for example, are so faint that they cannot be immediately discerned visually. But the more sensitive "eyes" of multispectral photographic devices see them, and separate and identify them.

Also under development is an advanced type of sensor designed to measure carbon monoxide concentrations in the atmosphere. It will be able to make global measurements of the poisonous gas, from space, enabling environmentalists to map those portions of the earth's atmosphere with high, low, and average concentrations of carbon monoxide.

NASA's overall program goal in pollution monitoring is to apply space technology, in cooperation with user agencies, to assessing the state of the environment and evaluating the future impact of candidate strategies for enhancing the quality of the environment. Fundamental objectives are the development of the techniques required to obtain the measurements of the important pollutants and the corresponding development of the ground-based aircraft and satellite systems from which to conduct these measurements.

Sensing devices, now carried on aircraft and eventually to be carried by satellite, to detect and determine the size of oil slicks have been devised by NASA specialists at the Ames Research Center in Mountain View, California. Measurements were made off the west coast of the United States from a light plane, and the sensors detected slicks resulting from heavy and light crude oils and light diesel oil.

In another project aimed at preserving ocean resources, scientists at NASA's Langley Research Center, Hampton, Virginia, are developing a lightweight precision sensor, more color-sensitive than the human eye, which can detect color gradation in water and enable oceanologists to spot pollution. From ocean-spanning spacecraft this thirteen-pound sensor—"Mocs," for multichannel ocean-color sensor—can scan a strip perpendicular to its line of flight, separating it into distinct blocks and identifying the color of each. Oceanologists can use this precise color information to detect pollution, locate areas where fish are likely to be, and study marine biology. For example, some shades of green show that plankton are present, which means that fish are probably feeding in the area, while many brown shades are indicators of pollution. Given consistent exact color information from Mocs, oceanologists can chart trends in pollution and marine life for use in preserving and developing ocean resources.

A number of other sensors, platforms, systems, and techniques are in various stages of development—among them: remote sensor models to measure the "total burden" of gaseous trace constituents in the troposphere; a gas filter correlation analyzer for carbon monoxide which can be modified to measure sulfur dioxide; a feasibility study for a satellite-borne gas filter analyzer that simultaneously could measure seven pollutants; a

Fourier transform interferometer to measure carbon monoxide and methane; laser radar techniques to determine sulfur dioxide concentrations in power plant emissions.

For detecting air pollution there are an experimental model of a small solid-state tuned cavity microwave spectrometer for measuring formaldehyde and neutron activation analysis techniques for measuring heavy atmospheric trace elements in industrial regions. Investigations into remote sensors of water pollution are directed primarily toward the measurement of chlorophyll concentrations, surface temperature, and the detection of oil slicks.

While these examples are representative of NASA's experimental work in the development of sensor techniques that could be used on spaceborne systems, other environmental protection programs are already being tested by satellite in earth orbit.

In one experiment involving ERTS-1, Dr. C. T. Wezernak of the University of Michigan is using information gathered from space to study the quality of water in Florida, Lake Michigan, Lake Erie, Southern California, and New York Harbor.

"In southeast Florida and in California we are concerned with ocean outfalls, effluent fields, and coastal processes," he explains. "Tampa Bay is a good example of an estuary and New York Harbor is a prime barge dumping area of chemicals and sewage sludge. And finally, the Great Lakes . . . where there is a great deal of public interest about water quality and water resources management. . . . We want to study the movements and dispersion of pollutants and maybe discover some we didn't know about.

"We need ERTS imagery so we can plan remedial action. In the case of Lake Erie, we have never had the large picture. . . . ERTS photos will help us map pollution in the lakes and should be useful for planning corrective action where necessary."

Strip mining, which has devastated parts of Kentucky, West Virginia, Pennsylvania, Ohio, and other mining states, decapitates mountains, guts farmlands, destroys wildlife habitats. Landscapes are denuded, and streams choked with silt because of erosion; rains and seepage add deadly acids. Coal strip mining operations in Ohio, for instance, have disrupted large areas and affected the quality of water in streams that drain from the lakes.

Mineralization increases astronomically in these streams: sulfate and iron levels soar and very few types of aquatic life can survive.

"One of the big problems," says Wayne A. Pettyjohn, an Ohio State University geologist and an ERTS investigator, "is that we don't know how much area on a day-to-day, month-to-month, or year-to-year basis has been stripped. In other cases, we don't know where some of the old mines are. ERTS-1 photographs will help us to map those areas which were mined many years ago, between 1935 and 1945, so we can evaluate the recovery of vegetation."

The initial experiment centers on five counties in eastern Ohio, and Pettyjohn says, "We hope to develop a pilot project for surveillance that can be used to monitor strip mining operations anywhere within the country within another two years."

In other early ecological applications of ERTS-1 findings more than 10,000 smokestacks, all pouring out particulate emissions, have been mapped in Virginia. Some of these pollution sources were unknown to state environmental protection agencies.

Preliminary analysis indicates that large wetlands and recreational areas can be mapped by ERTS with an accuracy as high as 99 percent. Turbidity variations have been imaged "very nicely" in Lake Ontario, Lake Champlain, Utah Lake and Kansas reservoirs.

From a brilliantly clear ERTS image of the New York City area Dr. Wezernak has pointed out that the tonal variations in New York Bight (the Atlantic Ocean) are due to the presence of suspended soil material and thus are indicative of water quality. In one photograph ERTS detected the remains of an iron acid dump made from a barge and revealed that the principal contaminants—sulfuric acid and ferrous sulfate—did not tend to dissipate very quickly. Further, repeated observations of the area indicated that the acid drift was southwesterly, resulting in damage to the New Jersey beaches.

Initial analyses of imagery from several successive ERTS-1 orbits showed the extent, predominant drift, and dispersion characteristics of waste disposal in coastal New Jersey waters. Other investigations have determined that within the bays, sounds, and thoroughfares behind the barrier islands in the southern New Jersey shore area the increased reflectance of the turbid waters illustrates the effect of a large sewage effluent flow. Because these waters are flushed with each tidal change, the turbidity reaches the area's populous bathing beaches.

From ERTS has come the first use of space imagery in the prosecution of an alleged violation of environmental laws. It was reported by Dr. A. O. Lind of the University of Vermont, who has been studying water pollution, lake turbidity patterns and land use in the Vermont area.

From an enlargement of an ERTS image, it was determined that the International Paper Company mill, located north of Ticonderoga on the New York side of Lake Champlain, was discharging treated wastes into the lake. The imagery, in conjunction with ground observation and measurements, determined the rate of discharge at 21 million gallons per day. The wastes contain suspended solids and are high in sodium and phosphates, which appear as reddish-brown in multispectral scanner imagery.

Based on this irrefutable evidence from space and on the corresponding ground data, the State of Vermont took legal action against the paper company and the State of New York, alleging that the plant was reducing the quality of the lake water to below Vermont standards, and pointing out that these pollutants cross state lines. The New York-Vermont state line runs down the middle of the lake.

"Actually, we're just starting in this field," says Ted George in a discussion of the use of spaceborne systems to monitor ecological conditions on the earth. "The day is coming when we will be able to map the entire United States and daily show pollutant contours, effects and concentrations, much as we now map weather patterns based on information relayed by satellite."

The ERTS projects concerned with the environment have thus far been more or less teasers. While their results have been dramatic and impressive, the experiments have covered only relatively small pockets of earth.

The goal is continuous coverage of earth from space on a truly worldwide scale. The next important step in assessing detection and monitoring of pollution from orbital vantage points, therefore, is to flight-test in a satellite the instruments that have been and are continuing to be developed in the laboratories and on aircraft. That is the specific objective of Nimbus-G, scheduled by NASA to be launched in 1977.

Speaking of the missions planned for this spacecraft, Charles Mathews says, "For the first time polluting gases and particles in our atmosphere will be measured globally from space. Temperatures and colors of the oceans will be more completely defined since precious measurements have been widely scattered and not closely coordinated.

"From these satellite measurements many of the pollutants in our lakes and oceans can be detected and tracked. This will involve determining the currents and roughness of the oceans. Oceanographic measurements, which are vital to the management of the world's fisheries and shipping, will be started with Nimbus-G.

The Nimbus-G satellite will be placed in a nearly polar orbit so that it views the earth repeatedly and each place is seen at the same local sun time, thus making possible more exact comparisons between the successive views.

Sensors planned for Nimbus-G are designed to measure gaseous pollutants at stratospheric levels and at tropospheric levels down to the surface. An instrument is also included to sense pollution particles to determine their size and density in the air.

Nimbus-G will also have oceanographic assignments because, like the atmosphere, the oceans are major sources and absorbers of heat, moisture, and pollutants. Oceanographic data from this satellite will be applicable to harvesting live sea resources, to routing shipping, and to reducing ocean pollutants, among other uses. Since the oceans' effect on climate governs the habitability of much of the world's land surface, such parameters as sea-surface temperature, ocean color, surface wind, sea-state, ice distribution, and currents will be measured by Nimbus-G instruments.

Another spacecraft system now on NASA drawing boards that will greatly contribute to the solution of the world's pollution problems is the Earth Observatory Satellite (EOS). Target date for launch is 1978, and the mission would be to mount an advanced orbital platform to make a number of specific measurements, including atmospheric pollution, cloud structure and composition, sea-surface phenomena, structure and phenomena of the atmosphere above twenty miles, and spectral, spatial, and temporal characteristics of significant earth and surface features.

"The EOS," says Mathews, "will continue, and improve upon, the terrain survey and weather and climate research previously provided by ERTS and Nimbus, but will initiate new programs in the study of oceanographic phenomena and environmental quality not now provided by these satellites.

"Studies have indicated a major overlap of weather and climate, and earth resources data requirements, which lead to the conclusion that future Nimbus and ERTS missions can best be combined into single earth observations missions."

Speculating on the effects of earth observation satellites, NASA administrator Dr. James C. Fletcher, in a speech to the National Wildlife Federation, prophesied: "Imagine a world fifteen or twenty years from now that has taken some advantage of the scientific knowledge and technical capabilities space exploration has already created.

"Think of a global weather network made up largely of observation satellite systems that constantly pour data into a computerized prediction system. This will let anyone have a 'nowcast' of the climate anywhere in the world as well as an up-to-date and accurate forecast valid for up to two weeks—also for anyplace in the world. The economic values of knowing—not guessing—the weather are enormous: they apply to agriculture, to construction, to transportation, to recreation.

"With the kind of understanding that such an information and prediction capability provides, it is not far-fetched to postulate operational, purposeful weather modification and climate control—on a local, regional, and global basis. The implications for human affairs are staggering—to eliminate the dangers of hurricanes, to relieve droughts, to ameliorate living conditions, to increase productivity, to balance imperiled ecologies—the list goes on.

"Take water," Dr. Fletcher continued: "By measuring the change in the snow pack, by knowing the weather, by having stream gauges that monitor flow, quality, temperature and report back to central facilities, by watching erosion effects and pollution sources and usage patterns, we can develop regional water management systems that combine the esthetic and the practical. "The same will be true for land use.

"Take the encompassing subject of environmental quality. Again, space systems observe, measure, and communicate, acting in conjunction with aerial and ground-based and buoyborne instruments. Information is instantly available—and to everyone. Are earthquake stresses building up? Minute motions and stresses are measured by automated monitors; the data are relayed by satellite; if a crisis is imminent, regional warnings are broadcast via space system; if not critical, there is time to relieve or redirect the growing stresses. Oceanic farming has replaced hit-or-miss fishing; we monitor sea-state, temperature, nutrients, and ocean quality as routinely as we now do the rainfall. Atmospheric quality is monitored by spaceborne instruments.

"No part of the changing, moving face of the globe we inhabit is free of human influence or removed from human interest. We therefore can afford to leave no part unmonitored. From forest fires to hurricanes, from the slow erosion of granite hills to the short-lived volcanic eruption or avalanche, from atmospheric particulates to crop diseases to oil spills—we need to know the condition of our environment, minute by minute. And from that kind of knowledge—perhaps *only* from that kind of knowledge—can and will flow the interrelated set of local, national and international measures that will make our planet what we want it to be.

"Everything I have suggested here is already possible," Dr. Fletcher said. "The technologies exist or are under development."

In summary, the mission of Nimbus-G and EOS are being designed to be directly applicable to significant environmental problems—among them air and water pollution—facing the nation and the world. In the instrumentation developed for these spacecraft man will have the tools necessary to monitor the world's air- and waterways effectively.

Even the finest information delivered by satellites, however, cannot control pollution. Such control must come not from space, but from ground—from enlightened attitudes, laws, and enforcement.

Nevertheless, without valid global data, the most important pollution problems cannot be determined, nor can the cost-effectiveness and success of methods of attacking them be established. Monitoring is obviously not a substitute for action. But action without the knowledge provided by adequate monitoring—including that from space—is likely to be ineffective.

Part Three

the
science

○
Chapter 9
the legacy of appollo

The Apollo flights demand that the word "impossible" be struck from the scientific dictionary. They are the greatest encouragement for the human spirit.

—ALBERT SZENT-GYÖRGYI

It has been called the most spectacularly successful series of explorations in history, the greatest engineering/scientific program of all time, the boldest, most challenging, complex, and profound human venture ever. But perhaps the most eloquent description was the simplest, expressed by astronaut Neil Armstrong when, at 10:56 P.M. Eastern Daylight Time, July 20, 1969, he became the first earthling to set foot upon another celestial body: "That's one small step for a man, one giant leap for mankind!"

The Apollo manned lunar landing program was all these—and more.

"No one can know where this exploration will finally take us," historian Arthur Schlesinger, Jr., writes. "But the pursuit of

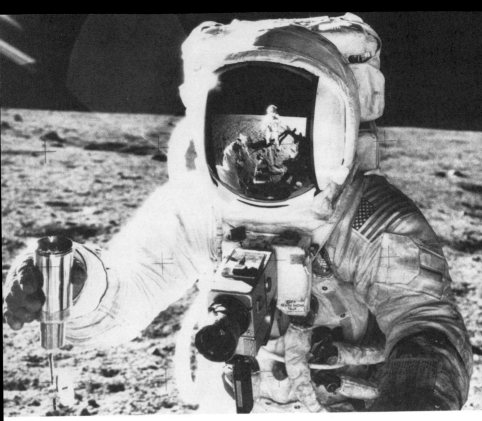

Mirrored in his teammate's helmet visor, an Apollo astronaut sets out experiments on the lunar floor. (Courtesy Space Division, Rockwell International)

knowledge and understanding has been human kind's most abiding quest, and to have confined this quest to our own small planet, to have refused the adventure of space, would surely have been a betrayal of man's innermost nature.

"The 20th century will be remembered, when all else about it is forgotten, as the century in which man first burst his terrestrial bonds and began the exploration of space."

By 1973 nine American teams of astronauts had made the epic journey to the vicinity of the moon, and twelve men had walked on the lunar surface. They brought more than 850 pounds of moon rocks and soil for analysis in earth-based laboratories, and established scientific stations that continue to transmit scientific and engineering data to earth from a quarter of a million miles in space.

The program, born on paper in July 1960, became a prime national commitment less than a year later with President John F. Kennedy's famous man-to-the-moon proclamation that sent this nation "sailing upon a new ocean."

The one-man Mercury series of space flights in the early 1960s and the two-man Gemini missions of the mid-60s gave astronauts and engineers the experience and confidence essential for the orderly, progressive march to the moon.

The following is a summary of Apollo flight highlights:

- *Apollo 7*, October 11–22, 1968: On the first manned mission in the program, the men—Wally Schirra, Donn Eisele, and Walt Cunningham—and the machines performed flawlessly for several days in earth orbit. Eight checkout firings of the spacecraft's primary propulsion system were made, and the first live television pictures were beamed from a manned vehicle in space.
- *Apollo 8*, December 21–27, 1968: Astronauts Frank Borman, James Lovell, and Bill Anders made history's first flight from earth to another body in the solar system when they circled the moon, traveling more than a half-million miles in space.
- *Apollo 9*, March 3–13, 1969: This mission, covering more than 6 million miles in earth orbit, included the first in-space checkout of the lunar module (LM)—the spidery-looking vehicle that was to take men down to the moon's surface. The crew were James McDivitt, David Scott, and Rusty Schweikart.
- *Apollo 10*, May 18–26, 1969: Spending nearly sixty-two hours in lunar orbit, Tom Stafford, Gene Cernan, and John Young flew this mission, which served as the full-scale dress rehearsal for the first manned lunar landing. It included separation of the lunar module from the Apollo command ship and the flight of the LM to within nine miles of the moon's surface.
- *Apollo 11*, July 16–24, 1969: Reaching the goal set eight years earlier, Neil Armstrong and Edwin Aldrin became the first men to explore the moon. (Mike Collins, the Apollo command module pilot, remained in lunar orbit.) Touching down in the Sea of Tranquillity, Armstrong uttered the words that will live forever in history: "The Eagle has landed." The Apollo 11 crew gathered an assortment of lunar rocks and was viewed live around the world by the largest television audience in history.

- *Apollo 12*, November 14–24, 1969: This second manned moon landing, in the Ocean of Storms area, included placement of the first Apollo lunar surface experiments package for continued science reporting. The crew was Charles Conrad, Alan Bean, and Dick Gordon.

- *Apollo 13*, April 11–17, 1970: A rupture of the service module oxygen tank, causing an electrical system power failure, cut short the mission objectives of astronauts James Lovell, Fred Haise, and Jack Swigert. The planned lunar landing had to be canceled.

- *Apollo 14*, January 31–February 9, 1971: The third manned lunar landing took place in the rugged Fra Mauro highlands, and included two lengthy explorations by Alan Shepard and Edgar Mitchell, while Stu Roosa orbited the moon in Apollo. Nearly 100 pounds of lunar rocks and soil samples were returned, and an array of scientific experiments set up on the moon's surface.

- *Apollo 15*, July 26–August 7, 1971: Modifications to the spacecraft permitted David Scott and James Irwin to spend nearly sixty-seven hours on the moon, eighteen and a half of them exploring the surface in the Hadley Apennine area. They retrieved approximately 170 pounds of samples, deployed geophysical instruments, and described geological features. Command module pilot Alfred Worden, in lunar orbit, conducted extensive scientific experiments.

- *Apollo 16*, April 16–27, 1972: The fifth lunar landing mission, to the Descartes highlands, included a record time of nearly three days on the moon's surface. John Young and Charles Duke had more than twenty hours for exploration and experimentation on the surface, while Tom Mattingly operated a complex array of scientific experiments in lunar orbit. Over 210 pounds of rocks and soil were returned to earth.

- *Apollo 17*, December 6–19, 1972: Gene Cernan and Dr. Harrison Schmitt, a geologist, made the final moon exploration in the Apollo program, while Ron Evans piloted the command module in lunar orbit. A number of major geological "finds" were made, including the discovery of "orange" soil, and nearly 250 pounds of samples were brought back to earth. A number of scientific experiments were placed about the Taurus-Littrow region on the moon.

What has the Apollo program accomplished? Dr. Rocco Petrone, associate NASA administrator, and formerly in charge of the Apollo program, has summed it up this way: "It has in-

*Dependable Apollo spacecraft—the fine
most complex transportation system ev
built by man—carried nine teams of astr
nauts to the vicinity of the moon ar
safely back to earth.*

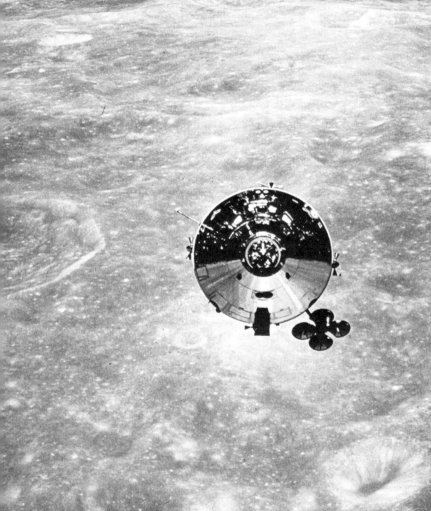

creased our knowledge of the moon beyond expectation. It has provided new knowledge and techniques for study of both the earth and sun. It has led to a wealth of new technology, and has given mankind a new frontier and the beginning to the technology necessary to exploit it.

"With the conclusion of the Apollo program, not only do we know much more about the moon, enabling us to formulate much more sophisticated questions about it, but equally important, we know more about the earth and the solar system as well. The moon is a scientific treasure house of knowledge to aid in understanding the origin and evolution of the terrestrial planets and the material they are made of.

"The Apollo program has proved that we can dedicate our resources to the achievement of a large-scale demanding technological goal without the stimulus of war."

Yet some of the Apollo program's most important returns could never be assessed or evaluated in scientific laboratories.

National pride is one example. In October 1957 the United States was stunned when the Russians orbited Sputnik I before any American spacecraft could get off the pad. Three and a half years later the American public was shocked again when, as Alan Shepard prepared for a short suborbital flight from Cape Canaveral, the Soviets orbited cosmonaut Yuri Gagarin, once again stealing a technological march on the world's greatest industrial nation. American prestige sank to new lows, and American engineering and scientific preeminence, long established and taken for granted, were openly and seriously threatened.

Apollo answered that challenge brilliantly, dramatically, and decisively. It was no secret that the Russians desperately wanted to land the first men on the moon, and through the 1960s they worked feverishly toward that goal—but so did the United States. Mobilizing a magnificent task force that at one time numbered 400,000 throughout NASA, other agencies of the federal government, and industry, the objective of landing men on the moon and returning them safely to earth was accomplished in spectacular fashion in July 1969—a little more than eight years after John Kennedy had set the goal. With the landing of Apollo 11 national pride ascended to new peaks, the astronauts were instant heroes, and America's preeminence in science, engineering, manufacturing, and other technological areas was firmly reestablished. There were more virtually immeasurable down-to-earth benefits, and while the program ended in December 1972, its effects are permanent and pervasive. Consider, for instance, the hundreds of thousands of jobs it created and sustained through the 1960s. Consider the schools and homes it caused to be built, and the nationwide economic boost.

Apollo also seemed to have a kind of international unifying effect, for as the astronauts sailed out in the spatial seas on each mission, explored the moon, and came home, people everywhere followed their voyages with intense interest. Hundreds of millions of earthlings—in North and South America, in Europe and Asia, in Australia and elsewhere—watched the proceedings live on television. The tensions and problems of everyday were, for these brief times at least, forgotten. The American astronauts, in a very real sense, were truly representing all the peoples of earth.

A plaque left on the moon by Apollo 11 astronauts Neil Armstrong and Edwin Aldrin reads: "Here America completed its first exploration of the moon. May the spirit of peace in which we came be reflected in the lives of all mankind." Commenting later on his flight, Armstrong said, "It was the first time in all history that explorers went to a new land without weapons of any sort."

"One of the most significant things . . . about Apollo is that it has opened for us—for us, meaning the world—a challenge of the future," says Apollo 17 commander Gene Cernan. "The door is now cracked, but the promise of that future lies in the young people—not just in America, but the young people all over the world—learning to live and learning to work together."

Equally as important was the realization of people everywhere, through Apollo, of how precious and fragile earth actually is. The British scientist and science-fiction author Fred Hoyle believes that one of the most significant fruits of space exploration was symbolized by the first photograph of earth, taken by Apollo astronauts from deep space. The psychological effect of this picture, Hoyle feels, was profound, for it dramatized the finiteness of the planet and its resources. And Hoyle dates the intense worldwide concern over environment, ecology, and depletion of resources from that photograph.

On this theme, Dr. Wernher von Braun, former NASA official and one of the Space Age's most respected pioneering leaders, has said, "Apollo has altered the concepts we had of ourselves, of our earth . . . to see earth as a complete and closed ecological system in the black of space was an emotional shock which shook us free of long-established, purely parochial concerns. I believe we can credit Apollo in particular, and the space program in general, for the surge in popular interest and anxiety for a cleaner environment, for better management of

Crescent earthrise across the pock-marked face of the moon was viewed by Apollo 17 astronauts.

natural resources, and for the maintenance of the ecological balance."

And what of the lift Apollo has given to the human spirit? How can this be measured? "If nothing else," Dr. von Braun adds, "Apollo has provided a new perspective and stimulated new lines of constructive thinking that enhance the brotherhood of man. It is the kind of impact on society, subtle but powerful, which moves individuals and nations toward a higher level of civilized conduct than previously seemed possible. The human spirit, when inspired, can accomplish miracles that no amount of satisfying our material needs can ever hope to equal."

Another view was presented by astronaut Frank Borman to a Joint Session of Congress following his moon-circling Apollo 8 flight: "Exploration really is the essence of the human spirit, and to pause, to falter, to turn our back on the quest for knowledge, is to perish."

With Apollo man cut forever his umbilical cord to earth. He proved convincingly that life can be sustained far from his home planet. Will history record this step as being as significant as that of the first amphibian who crawled from the sea onto the land and began to conquer a new world?

Time will tell.

○
Chapter 10

lunar gemstones

This is a geologist's paradise if I ever saw one.

—ASTRONAUT HARRISON SCHMITT
ON THE SURFACE OF THE MOON

While he has undoubtedly wondered about it as long as he has been on earth, yet it is less than four hundred years—Galileo began observing it with the first telescope in 1609—since man has been able to begin a book of knowledge about the moon.

It was Galileo, for example, who noted that the lunar surface visible from earth consisted of mountainous regions, which he designated *terra,* and of smoother, flatter regions, which he labeled *mare*—by analogy to the terrestrial oceans and continents. He further perceived that there was a marked difference in the reflectivity of these two: the *mare* areas were much darker than the *terra.*

Astronomical studies in the less than four centuries since

then have added a good deal to the book. But because all the observations were made from earth—roughly a quarter of a million miles away—the information assembled was limited to the lunar surface features visible from earth. Even the most sophisticated, sensitive telescopic instruments cannot provide the variety or detail of scientific data that are necessary to the understanding of the history and evolution of the moon.

The Age of Apollo has changed that. Incredibly, man has learned more about his planet's only natural satellite since July 1969, when Neil Armstrong and Edwin Aldrin first set foot upon the lunar surface, than he had in all previous millennia.

Yet, why are we interested at all? What does a seemingly lifeless body so far away have to do with life on earth?

"The moon affords us a unique opportunity," says Dr. Rocco Petrone. "It is the only place in the solar system where there is a permanent record preserved. Earth and Mars have changed since their formation. It is likely other planets have too. But the moon has not gone through these evolutionary stages. Therefore, in studying it, we can look back at four and a half billion years of undisturbed history.

"From this, we can gain an insight into our own being. Man must understand the total relationship with the sun, and by learning about the moon, we are doing that. By exploring our nearest neighbor, we're starting to unravel how the process of planet formation was accomplished. We're fundamentally looking back into the time of the creation. . . .

"Man has looked at the moon ever since he has been on earth. Now he has pieces of it in his laboratories. No one lunar rock will give us the history of the moon, or answer all our questions. But these rocks are building blocks, pieces in the puzzle."

Scientists believe that what we will learn from the Apollo program about the moon and its relationship with earth, the sun, and the solar system could have as great and profound an impact on the intellectual and ordinary life of man as did the theories of a sun-centered planetary system, human evolution, and of relativity.

The key questions to be answered are:

- When, where, and how did the moon originate?
- What history and geological features do the moon and earth have in common, and what are the differences?
- What exactly is the moon made of?
- How is its subsurface structured?
- What is its crust like?
- What is the moon's magnetic field?
- What can the moon tell us about earth, the sun, the other planets in the solar system, and the universe?
- Is there any evidence of past or present life on the moon?

Scientists have long debated these issues, but until Apollo there was far more theory than proven fact, more hypothesis and speculation than hard knowledge. And the educated guesses ranged widely, from the routinely predictable to the bizarre.

Apollo began providing answers from the instant that Neil Armstrong set foot upon the moon: "The surface is fine and powdery," he reported, seconds after he had descended from his lunar module. "I only go in a small fraction of an inch, maybe an eighth of an inch, but I can see the footprints in the fine sandy particles. There seems to be no difficulty in moving around, as we suspected." Later, he added, "It's a very soft surface, but here and there . . . I run into a very hard surface, but it appears to be very cohesive material of the same sort. . . . It has a stark beauty like much of the high desert of the United States."

Scientists clung to every word, spellbound by the first-hand observations, for they realized that Apollo and the United States space program were fostering the birth of a new field—Lunar Science, which encompasses geophysics, geochemistry, bio-science, geodesy, cartography, and atmospheric and particles studies.

During their historic visit at Tranquillity Base, Armstrong and Aldrin gathered handfuls of lunar rocks and scoops of soil, carefully labeled individual packages, and placed it all—17.6 pounds—in specially sealed containers. This cargo was probably the most precious mineral man has ever known. Scientifically, at least, it was far more valuable than its weight equivalent in gold, silver, platinum, diamonds, emeralds, or any other matter on which the highest earthly value is placed.

Microscopic view of moon soil brough back by the Apollo 11 crew reveal glassy spherules.

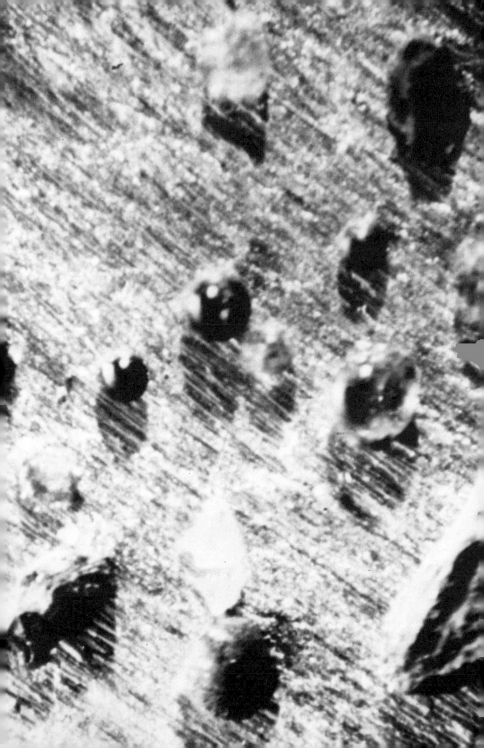

Why are the lunar rocks so important? The shape, size, arrangement, and composition of the individual grains in a rock tell scientists about its history. Radioactive "clocks"* tell them its age, and tiny tracks can reveal the sun's radiation history during the last 100,000 years. Called by some the Rosetta Stones† of the solar system, the rocks went through a series of microscopic examinations, measurements, and weighings. Cameras photographed samples from every conceivable angle. Two tiny chips or pinches of dust were taken from each specimen. They were run through the physical-chemical areas of the laboratory for fairly detailed evaluation of their mineralogical, petrological, and geochemical compositions.

Some of the investigative work was done quickly after receipt of the samples, including such "time critical" experiments as the determination of cosmic ray and natural radioactivities in the material—information that would disappear if not obtained soon after the samples reached earth.

"Our purpose is to understand the material, to carry out a series of physical, chemical, and mineralogical analyses that will enable us to understand the contents of the samples," explained Dr. Wilmot Hess, chairman of NASA's preliminary examination team at the time of Apollo 11. The team helped laboratory scientists catalog, sort, describe, and test the material.

Bits, chips, and tiny fragments and particles of the moon were released to 142 principal investigators, or PIs, located at major universities, laboratories, and research centers around the world. Never in history had such an internationally renowned scientific team been assembled to participate in one project, each person a distinguished authority, each acknowledged as a world

---

* One of the primary means of determining a rock's age is reading its "atomic clocks." For instance, one isotope of the element potassium, found in the lunar samples, is radioactive. This decays at a known rate and becomes an inert isotope of the gas argon. By determining the ratio of "parent" potassium to "daughter" argon, it is possible to read the age of the sample.
† The Rosetta Stone, a stone slab found in 1799 near Rosetta on the Nile River, bears parallel inscriptions in Greek, Egyptian hieroglyphics, and demotic characters, making it possible to decipher ancient Egyptian hieroglyphics.

leader in his field of study, several Nobel Prize-winners in physics and chemistry.

Depending upon individual requirements, the PIs were allotted samples ranging from about 20 milligrams (barely a visible speck), to 250 or more grams (about a half-pound). During the next several months these bits of the moon were inspected, examined, isolated, crushed, sliced, hammered, burned, powdered, disected, exposed to radioactivity, and subjected to a multitude of other scientific torture tests—all in a relentless effort to extract every iota of meaningful data.

Scientists used a variety of analytical tools—computers, mass spectrometers, X-ray diffraction and fluorescence units. The investigations covered mineralogy and petrology, crystallography, microprobe analysis, radiation and shock effects, alpha-particle autoradiography, chemical, isotope, and rare gas analysis, cosmic ray-induced and natural radioactivity measurements, light-stable isotope, magnetic, thermal, and elastic-mechanical measurements, determination of electrical and electromagnetic properties, and biochemical and organic analysis.

The same basic procedure for the processing and examina-

Priceless gem: This unusually large rock, brought back by astronauts from the surface of the moon, is prepared for initial examination at the Lunar Receiving Laboratory at NASA's Johnson Space Center, Houston, Texas.

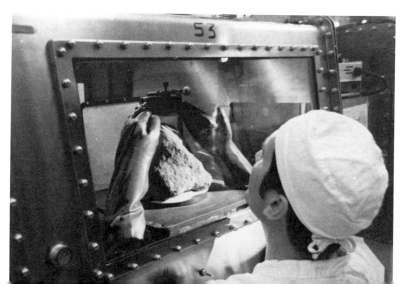

tion of lunar rocks has been followed throughout the Apollo program. All samples were initially inspected, catalogued, and run through preliminary analysis at NASA's Lunar Receiving Laboratory; select portions were then assigned to principal investigators for more detailed, specialized study.

Early in January 1970 the Apollo 11 PIs and other interested scientists—more than 850 from all over the world—assembled at Houston to discuss their findings at the first Lunar Science Conference. The results of their analyses generated wide-ranging interest and in some cases astonishment that has not yet substantially subsided. Dr. Werner von Braun said, "The moon has proved to be a wonderland of surprises."

The oldest material found at Tranquillity Base was estimated to be about 4.4 billion years old. This in itself was of major significance, for the oldest known rocks on earth are granitelike specimens a little more than 3.7 billion years old found in western Greenland.

In May 1970 scientists announced that a component of one of the rocks returned by the Apollo 12 crew had indicated, in laboratory analysis, that it was about 4.6 billion years old. The dating was especially significant because it corresponded with previous scientific dating of samples from elsewhere in the solar system, particularly meteorites.

Many astronomers and geochemists believe that 4.6 billion years ago was when the planets and the sun were formed. This lemon-sized rock designated no. 12013, light gray with cloudy-white crystals and dark-gray streaks seems to indicate that the moon may have existed, with its existence relatively undisturbed, millions of years longer than previously thought.

In tying in the age of the moon to other elements in the solar system, many scientists agree that something cataclysmic happened 4.6 billion years ago. It could well have been the final coalescence of the gigantic dust and gas cloud from which sun, planets, and moons were born. There are no definite answers as to the origins of the gas.

It must be pointed out, however, that the apparent evidence from this single rock was not, as NASA has described it, "compelling." A determined search for identifiable "old" rocks was made on each lunar landing, and one found on a later mission, no.

Scientist-astronaut Dr. Harrison Schmitt sets out an experiment on the moon next to a huge split lunar boulder during the Apollo 17 mission.

65015, strengthened the belief that rocks that old did in fact exist on the lunar surface. It was a partially recrystallized breccia—a rock containing another rock—in which, apparently, the unrecrystallized pieces could be estimated to be from 4.4 to 4.5 billion years old. Breccias on the moon are formed under the intense pressure and temperature produced by meteorite impacts.

The lunar samples are unquestionably of unprecedented scientific value, for they contain a record of the birth and growing pains of a planet—a record that can never be read in earth rocks because continual geological activity, such as crustal upheavals and weathering by wind and water, has destroyed earth's oldest rocks.

Summing up the findings from the early Apollo lunar landings, Dr. John Wood, a principal investigator and staff scientist at the Smithsonian Astrophysical Observatory, said, "From a relative handful of lunar material and some sophisticated observations, we can see a moon being born in violence and intense heat, cooling later to form its thick rocky crust. During the next billion years, this crust would be violated from below as molten lavas forced their way to the surface, and bludgeoned from above by a rain of planetary debris. Then, in some age yet to be determined, the moon died, geologically, and became a place of few significant events until men arrived."

While the most detailed and varied data obtained on the moon involve the surface materials—rocks and soil—brought back by six teams of astronauts, a hoard of other information continues to be gathered for earthbound analysis and interpretation by an extensive scientific network placed on the moon by the Apollo crews.

This includes five independent experimental stations, a $125-million nuclear-powered system that allows direct relay of data from the moon to scientists on earth. Instruments in the network measure such phenomena as tremors beneath the surface, heat radiating from within, gravitational and magnetic forces, and particles in the atmosphere.

All five stations include seismometers to record tremors caused by meteoroids, tidal stresses, and internal changes in the moon. Because the moon is extremely quiet, the seismometers register shocks far smaller than would be noticeable on earth. The seismic network is providing a wealth of information about layers far beneath the lunar surface. Scientists have learned that the moon has a thick, solid crust and perhaps a molten core.

Even though the lunar atmosphere is a nearly perfect vacuum, various instruments measure the constant changes—solar winds, gases escaping from below the surface, and dust thrown up by the impact of meteroids—all of which contribute to matter in the atmosphere. Special sensors, placed in holes drilled eight and a half feet into the lunar crust, are designed to measure the heat emanating from the center of the moon, while other experiments have been set up at different stations to measure lunar gravity, the size and speed of micrometeroids, and the elements of the lunar atmosphere.

Additionally, three arrays of specially constructed laser-beam retroreflectors were sent from observatories on earth to be placed on the moon during the Apollo 11, 14, and 15 missions. By aiming a laser beam at any of the arrays and timing its return to earth, the distance to the moon can be calculated with unprecedented precision—to within six inches.

Other key experiments were conducted above the lunar surface. Before leaving lunar orbit, the Apollo 15 crew launched an 81-pound scientific "subsatellite." It carried three experiments: one to study the lunar gravity fields, the other two to collect data on the moon's magnetic field and its interaction with the solar wind. A second subsatellite was launched by the Apollo 16 crew, and both orbiting craft have provided scientists for the first time with magnetic data from the far side of the moon and new data on lunar gravitational characteristics.

As confidence in Apollo increased, the crews stayed longer and longer on the lunar surface. From the Apollo 15 mission on, they greatly increased their mobility on the moon by using specially designed lunar roving buggies.

But perhaps the most significant advance during the series of flights was the presence, on the Apollo 17 crew, of the first real scientist to explore space. Dr. Harrison "Jack" Schmitt, a geologist, has a Ph.D. from Harvard, was a Fulbright Scholar, and made immeasurable contributions to the program.

It was Schmitt and Cernan who discovered the exciting "bright-orange" soil, which suggests that the area where it was found may contain relatively young volcanic features. The material appears to be a finely structured glass of volcanic origin.

Schmitt's incisive disciplined comments and observations throughout the Apollo 17's exploration of the Taurus Littrow area greatly aided scientists in their studies on earth. Apollo 17 demonstrated convincingly the great contributions a trained scientist can make in space.

From six manned lunar landings—Apollos 11, 12, 14, 15, 16, and 17, between July 1969 and December 1972—an enormous mass of data have been accumulated. A total of 841 pounds of moon rocks and soil samples, from six separate sites ranging from the low mare areas to the lunar highlands, has been returned for study.

The twelve astronauts who physically explored the moon, spent about 160 man-hours, traveling more than 60 miles afoot and by lunar roving vehicles. NASA estimates as many as 30,000 photographs from Apollo alone have captured the moon in intimate detail, and nearly 90 major scientific experiments were performed throughout the program, 60 on the surface and another 30 in lunar orbit.

Many of these facts and observations have already been tentatively assembled into theories and models that are leading toward a genuine understanding of the moon's history. In other cases more study is needed, more data must be correlated, before even educated assumptions can be made. Answers to some questions have merely led to more complicated questions. Scientists will be analyzing and evaluating this great storehouse of information for years, probably decades.

From all the data collected and studied, a more detailed picture of lunar history has already emerged. We know now that the moon is a complex body whose origin dates back to the beginning of the solar system; that its formation and evolution can be related to the more general planetary processes; and that a record of events in the history of our sun and our galaxy may be found on moon's surface.

NASA has issued the following more detailed statement on major areas of understanding directly produced by the Apollo program:

- "We now have a rather definite and reliable time scale for the sequence of events in lunar history. In particular, it has been established with some confidence that the filling of the mare basins largely took place between 3.1 and 3.8 billion years ago. Since these surfaces represent the major physiographic features on the lunar surface, we can immediately infer that the bulk of lunar history recorded on the surface of the moon—that is, the time of formation of more than 90 percent of the craters—took place more than four billion years ago." (The spectacular Copernicus crater apparently was gouged out 850 to 950 million years ago, making it one of the more recent major events on the moon.)

"This is quite different from the terrestrial situation, where most of the earth's ocean basins are younger than 300 million years, and rocks more than 3 billion years older make up an almost insignificant proportion of the earth's surface.

- "The relative importance of volcanic and impact-produced features on the lunar surface is today rather well established. There is almost unanimous agreement that the dark mare regions are, indeed, underlain by extensive lava flows . . . On the other hand, almost all craters appear to be caused by impacting projectiles.

- "Compared to the earth, the moon is seismically very quiet. This is, of course, consistent with the conclusion that volcanism and other types of tectonic activity have been rare or absent from the lunar scene for the last 2–3 billion years. We have learned . . . that the moon has a crust more than 40 miles thick. . . .

- "We now have a much more detailed understanding of the moon's present magnetic field. It is clearly not negligible as we thought prior to the Apollo missions. Magnetometers placed on the lunar surface reveal a surprisingly strong, but variable field. (This is said to suggest that the moon has a metallic core as much as 175 miles thick.)

- "The fluctuation in the magnetic fields measured at the lunar surface is a function of the flux of incoming charged particles (solar wind) and the internal electrical conductivity of the moon. Careful study of these fluctuations shows that the moon has a relatively low conductivity.

- "The heat escaping by conduction from the interior of a planet depends on the amount of heat produced by the decay of radioactive elements, the thermal conductivity of the deep interior, and the initial temperature of the deep interior. Using rather imprecise models and making reasonable assumptions concerning the abundance of the radioactive elements potassium, uranium, and thorium, it was expected that the energy flux from the interior of the moon would be substantially lower than that for the earth—simply because the moon is much smaller.

"The first measurement of this quantity at the Apollo 15 site indicates that this is not the case. If this measurement is characteristic of the whole moon, the only plausible explanation that has been put forth to date requires: First, that the moon is richer in the radioactive elements uranium and thorium than the earth; and secondly, that these elements are strongly concentrated into the upper parts of the moon.

"When combined with the observations on the volcanic history of the moon and the present-day internal temperatures, the energy flux leads to two current pictures of lunar evolution. The first assumes that the variation in radioactivity with depth is a primary characteristic of the planet; that is, the planet was chemically layered during its formation. In this case, the initial temperature of the lunar interior below 300 miles was relatively low, and the deep interior of the moon gradually became hotter, perhaps reaching the melting point during the last billion years. . . .

• "The most extensive and diverse data obtained on the lunar surface are those concerned with the chemistry and mineralogy of the surface materials. The study of samples from the six Apollo sites . . . reveals a number of chemical characteristics that are apparently moonwide. There is, nevertheless, some hesitancy to generalize from these relatively minute samples to the whole lunar surface.

"Fortunately, two experiments carried out in lunar orbit provided excellent data regarding the regional distribution of various rock types. The X-ray fluorescence experiment very convincingly defined the prime difference between the chemistry of the mare and highland regions. It showed that the highland regions are unusually rich in aluminum—much richer, in fact, than most terrestrial continents.

"This (and other) observations lead to the strong hypothesis that the regolith (rocks) and soil of highland regions is underlain by a 'crust' similar to the terrestrial rock designated anorthosite. . . .

"Both the samples and orbital geochemical experiments indicated that the three most common rocks in the lunar surface are plagioclase—or aluminum-rich anorthosites; uranium-, thorium-rich 'KREEP' basaltic rocks; and iron-rich mare basalts." (KREEP, whose name is an acronym using the chemical abbreviations for potassium, rare earth elements, and phosphorus, is a highly radioactive material rich in these elements; its origin remains a mystery.)

"The differences between the lunar rocks and terrestrial rocks are so marked that we can conclude that the moon must be chemically different from the earth."

In some cases, confident answers can be extracted from the Apollo data. The question of life on the moon is one good example, for no chemical evidence that living things (except the astronauts) has been found—no fossils, no microorganisms, no traces of biologically formed chemicals, nothing.

But whether the findings mean that certain scientific debates will rage on, or that some theories can be weeded out and others studied further, or that some data can be fully accepted, there is one point upon which all agree: because of the six manned American landings on the lunar surface, more has been learned of the moon's early history than the sum of knowledge accumulated thus far on the comparable early history of the earth.

And we have amassed this much knowledge, for the most part, since mid-1969. What can we expect from future efforts? Hundreds of scientists and specialists in the United States and a number of foreign countries are continuing exhaustive studies on the lunar samples. Somewhat surprisingly, less than 10 percent of the total 841 pounds of moon rocks and soil brought back to earth have been used for analysis; the remainder will be carefully preserved for future scientific investigation, probably using more powerful analytical tools not even known today.

NASA officials say that the backlog of detailed study on the moon samples parceled out to principal investigators will carry well into 1975, and perhaps beyond. All told, completion of scientific work, including analysis of data from the automated lunar stations, could take another five to seven years. At the splashdown of Apollo 17 in December 1972, NASA predicted that live reports from the network of stations at various sites on the moon would continue at least for another two years. But that is just their "advertised" lifetime. Most of these experiments had a life design goal of one year, but four of the five experiments carried to the moon on Apollo 12, in November 1969, were still operating and returning information to earth more than three years later. Final analysis and evaluation of these data will extend years beyond the day they finally cease to function. In fact, NASA's chief lunar scientist, Noel Hinners, has said, "We are looking forward essentially to as much as a decade of continued work on the data that is on hand."

"Encouraging and impressive as our results have been to date, they represent only a start on the analysis and assimilation of data," says John E. Naugle, deputy associate NASA administrator. "During the next decade, our efforts must be devoted to the systematic analysis and interpretation of the growing storehouse of data."

While it will take years to sort and analyze, evaluate and interpret, draw conclusions and develop theories and hypotheses, NASA also has at least tentative plans to continue to mine the mother lode of data still stored on the moon. The agency hopes, for instance, to inaugurate a program to develop unmanned moon-orbiting spacecraft. These would map the total lunar surface, study the moon's gravity, magnetic fields, and surface chemical characteristics, and gather information on the hidden backside.

Ultimately man will colonize the moon. It may or may not happen this century, but it will happen. It is inevitable in the natural progression of exploration.

Because of the advanced state of technology and its continued growth, it probably will be easier to establish permanent of semipermanent colonies on the moon than it was to sustain life in America early in the seventeenth century, when it took weeks merely to transport settlers across the ocean. We can fly to the moon in three days, and moon colonists will not have to face the ravages of disease or the threat of wild animals or savage natives, as America's colonists did. Of course, the lunar environment is much more alien to life, but astronauts have proven that technology can be used to overcome a hostile environment. All the supplies necessary to sustain a base on the moon, including water if necessary, can be ferried via spaceships from earth.

Why will man one day colonize the moon?

"We have only scratched its surface," says Dr. Petrone. "We must exploit our capability to send explorers, manned or unmanned, to the moon to increase knowledge. Apollo won't answer all man wants to know. It is the first phase of manned lunar exploration. It sets a good basis of understanding. Sometime later, when it is appropriate, perhaps we will go into the second phase. We will man bases on the moon.

"The question really is, then, why increase knowledge? Why did Plato and Socrates ask why? What was Newton after when he asked why? They believed the development of knowledge is the meaning of man. We believe today one of the fundamental questions man must pursue is what is his destiny. Why is he here? We must continue this quest for man to progress."

○
Chapter 11

astronomical wonders

To the average person, probably the least understood part of America's multi-faceted space program includes physics and astronomy. Yet in terms of scientific interest, this area promises greatly to advance man's scant knowledge of little-known physical phenomena that directly affect life on earth. This includes the highly complex magnetosphere surrounding the planet, and the effect of solar radiation on the ionosphere and atmosphere. Further out, spacecraft orbiting above the obscuring curtain of earth's atmosphere have made detailed undistorted observations deep into space to study ultraviolet, infrared, X-ray and gamma ray radiations to learn more about the stars, galaxies, pulsars, and quasars.

Additionally, hundreds of balloons and small sounding

rockets are launched from a variety of ranges throughout the world in projects designed for the study of near-earth space phenomena.

This varied assortment of probes—gathering data from barely a few miles above earth to distant points in star systems billions of miles away—are all aimed toward the major objective of increasing knowledge and understanding of the cosmic environment.

Man's picture of himself and his relationships to the universe influence his actions. This has been evidenced during the last few years in that the space program has led to a much better appreciation of our immediate environment. The direct practical contributions will be no less important. To understand changes in our environment and in the forces that shape it is vital.

Space technology is important to science for a very fundamental reason—a scientist can place his instruments in previously inaccessible locations—and so gain knowledge that was unavailable as long as he and his instruments were confined to the surface of the earth. The knowledge that we gain by placing our instruments in space is manifold. We will better understand how energy is produced and converted into its various forms, and the processes which control our weather patterns and our climate. We will certainly gain new insights into the nature of the universe and the laws which govern it.

What are some of the basic questions physics and astronomy missions may answer about the physical phenomena surrounding the earth and the sun and in deep space?

- Studies of the upper atmosphere will identify the roles played by naturally occurring molecules such as nitric oxide and ozone, and define any danger from potential contaminants.
- Through solar experiments, man will gain a better understanding of sun-earth relations which will help answer other questions about changes in long-term weather and climate patterns.

Solar studies are of special significance because so much of what happens on the sun directly affects earth. Also, more information about superhot plasmas, which man can study only in the sun, may contribute to its eventual applications on earth. For example, properly understood and harnessed, such plasmas could help solve continuing energy needs created by the gradual exhaustion of the fossil fuels—gas, oil, and coal.

• From instrumented astronomical observatories launched into space, we will gain an understanding of various processes that produce energy. We know that there is regular energy production in a magnetized hot ionized gas in solar flares. What particularly excites scientists is the release of energy in X-ray stars and galaxies on an unimaginably larger scale.

From vantage points on earth man has only an extremely limited view of a universe that can be "seen" in infrared, ultraviolet, X-ray, and other wavelengths of the electromagnetic spectrum.

Infrared light carries information about dust clouds and cool bodies, and permits observation further toward the center of "our" galaxy, the Milky Way. Ultraviolet light is emitted primarily by hot stars and steller coronae. The absorption of this light permits the observation of interstellar molecules and dust. X-rays and gamma rays are emitted by highly energetic processes that lead to the production of enormous amounts of energy such as that emitted by unusual astronomical objects called quasars and pulsars. Galactic cosmic rays constitute the only matter reaching earth from outside the solar system. They provide information on many sources in the galaxy.

To learn more about all these phenomena, man must extend his "eyes" by placing observation platforms in space, free from the distorting effects of earth's atmosphere. The advantages of having a telescope in orbit were dramatically exemplified in May 1972, when a star exploded in a nearby galaxy, producing the brightest supernova in thirty-five years. The first observations in ultraviolet of such an event were made by NASA's second Orbiting Astronomical Observatory—OAO 2. It showed that not only the interior but even the surface of an exploding star must be exceedingly hot.

OAO 2 made many important contributions to astronomy during a four-year lifetime in which it orbited the earth 22,000 times. It mapped a star atlas in the ultraviolet spectrum and analyzed the light emitted by a large number of stars. It found many more very hot stars in other galaxies than earthbound observers had predicted. It showed that ultraviolet observations are essential for understanding comets. And it made major contributions to the understanding of unusual features in the light emitted by several types of peculiar stars.

OAO 2 was retired in February 1973, but its work has been continued and expanded upon by OAO 3, called Copernicus to commemorate the five hundredth anniversary of the birth of the Polish genius who was the father of modern astronomy.

The largest (4900 pounds) and most complex unmanned spacecraft ever launched by the United States, Copernicus has made thousands of observations of ultraviolet and X-ray sources in the sky. Preliminary scientific results attained from a 32-inch reflecting telescope include detection of large quantities of molecular hydrogen in the denser interstellar dust clouds, observation of surprisingly large amounts of deuterium (a heavy form of hydrogen) in interstellar dust clouds, and determination that some solid particles or dust grains in interstellar clouds are smaller than believed previously—some less than one-millionth of an inch in diameter.

Scientists will probably be analyzing data returned from experiments aboard Copernicus for years. While it is the last large orbiting astronomical observatory in NASA's program, future flights are planned for other smaller satellites.

One of the essential elements of NASA's solar research program is the Apollo Telescope Mount (ATM) on Skylab. Its instrumentation is larger and more sensitive than any previously flown in space.

The following is a summary of major space physics and astronomy projects planned for the 1970s.

Atmosphere Explorers are 1320-pound spacecraft launched into elliptical earth orbit of different inclinations to make scientific studies of the photochemical processes and energy transfer mechanisms which control the structure and behavior of the earth's atmosphere and ionosphere through the region of high solar energy absorption. To date four of the spacecraft have been orbited, and two more are planned for 1975 launchings.

A small (88-pound) *Dual Air Density Explorer* will obtain global density information on the upper thermosphere and the lower exosphere, to measure the vertical structure of the atmosphere as a function of latitude, season, and local solar time. It also will perform composition measurements—with a unique mass spectrometer system—of the upper atmosphere and the lower exosphere. Its launch into earth orbit is set for late 1975.

In 1977 and 1978 three *International Magnetosphere Explorers*—a joint effort between NASA and the European Space Research Organization (ESRO)—will be launched into highly elliptical earth orbits. This mission will investigate solar-terrestrial relationships at the outermost boundaries of the earth's magnetosphere, to examine in detail the structure of the solar wind near earth and the shock wave which forms the interface between the solar wind and earth. Additionally, the IME spacecraft will continue the investigation of cosmic rays and solar flares in interplanetary regions.

A 1416-pound *International Ultraviolet Explorer* is being designed to obtain high-resolution ultraviolet data on the spectra of many types and classes of astronomical objectives. It will carry an 18-inch telescope that will be able to look ten times further into space than observation equipment aboard the Copernicus spacecraft. It will be launched in 1976 as a cooperative venture of NASA, the United Kingdom, and ESRO.

The highly successful *Interplanetary Monitoring Platform* project concluded a series of ten launches late in 1973. These spacecraft were designed to extend knowledge of sun-earth-moon relationships by conducting a continuing study of the radiation environment of the interplanetary medium. They perform detailed and near-continuous studies of the interplanetary environment for orbit periods comparable to several rotations of active regions of the sun. They also study particle and field interactions; investigate, during a period of changing solar activity through several solar rotations, the nature and features of the solar wind, the interplanetary field, and cosmic rays.

An *Orbiting Solar Observatory*, OSO 1, scheduled for launch in mid-1974, will investigate the sun's lower corona and the chromosphere and their interface in the X-ray and ultraviolet spectra, to obtain a better understanding of the transport of energy from the photosphere into the corona. Also to be studied are solar-terrestrial relationships and the background component of cosmic X-rays.

Launched in June 1973, a unique *Radio Astronomy Explorer* probed the sources of little-understood low-frequency signals from galactic and extragalactic radio sources, and from the sun, earth, and Jupiter. When fully extended in space, its antennas were as tall as the Empire State Building.

A *Small Astronomy Satellite,* to be launched in 1975, will survey the celestial sphere and search for sources radiating in the X-ray, gamma-ray, ultraviolet, and other spectral regions inside and outside earth's galaxy.

An earlier satellite in this series—SAS-B, or Explorer 48—observed the little-understood gamma radiation in our galaxy and beyond for a six-month period in the first half of 1973. Later in the decade, probably between 1977 and 1979, NASA plans to launch High Energy Astronomy Observatories (HEAO) to study some of the most intriguing mysteries of the universe—pulsars, black holes, neutron stars, and supernovas. We know little about these phenomena. Pulsars emit extremely precise radio signals, and most available evidence suggests that they may be fast-spinning neutron stars. These compact bodies of densely packed neutrons (atomic particles with no electric charge) are believed to form when a large star burns up its fuel and collapses. Containing the mass of a star in a sphere ten miles in diameter, a pulsar is so closely packed that a single spoonful of material from its center, incredibly, would weigh a billion tons.

Black holes are believed to be the final stage in the collapse of a dying star. A black hole's material is so densely packed—even more so than that of a neutron star—and its gravitational force is so great that even light waves are unable to escape from its surface.

The nature of quasars still baffles astronomers, but many now believe that among observable objects they are the most remote in the universe. They look more like stars when viewed through an optical telescope, but emit more energy at radio frequencies than the most powerful galaxies known. According to calculations, if they are as distant as many astronomers think they are, the total amount of energy emitted by a quasar in one second would supply all of earth's electrical energy needs for a billion years!

Mating: Radio Astronomy Explorer satellite is joined to Delta rocket at Cape Kennedy prior to launch. This spacecraft is designed to probe the sources of little-understood, low-frequency radio noise in our galaxy and beyond.

Radio galaxies, located on the fringes of visibility, emit radio waves millions of times more powerful than the emissions of a normal spiral galaxy, but no one knows what they are. Several of them broadcast with such power that a sizable fraction of the nuclear energy locked up in their matter must be going completely into the production of radio waves.

A supernova is the life's end of a large star whose final collapse is a cataclysmic event generating a violent explosion that blows the innards of the star into space where the materials mix with the primeval hydrogen of the universe. Later in the history of the galaxy, other stars are formed from this mixture. The sun, which is one of these stars, contains the debris of countless others that exploded before it was born.

HEAO spacecraft will permit the exploration of previously inaccessible sources of celestial X-rays, gamma rays, and cosmic rays. These observations should make a major impact on man's understanding of the newly discovered energy processes and the creation of matter.

NASA is also proceeding with detailed planning of a large multipurpose optical telescope to be launched and serviced by the space shuttle in the 1980s. Called the Large Space Telescope, or LST, it will be able to look at galaxies a hundred times fainter than those visible through the most powerful ground-based optical telescopes. And as an aid to planetary scientists it will be able to provide long-term monitoring of atmospheric phenomena on Venus, Mars, Jupiter, and Saturn.

Astronomers expect the LST to contribute significantly to studies relevant to the origin of the universe. It may, for instance, be able to answer such mind-stretching questions as how far the universe extends in time and distance. It may make crucial contributions to the understanding of the content, structure, scale, and evolution of the universe.

The LST will weigh between 20,000 and 25,000 pounds, have a length of 40 to 52 feet, and a diameter of 12 to 13 feet. Its diffraction-limited mirror alone will be approximately 10 feet in diameter.

Among its assignments will be the studies of the energy processes that occur in galactic nuclei and of the early stages of stellar and solar system formation, and observation of such highly evolved objects as supernova remnants.

In support of these far-ranging spacecraft programs, NASA annually launches about eighty sounding rockets from sites in North and South America, Europe, Asia, and Africa. Sounding rockets are launched for short-term measurements to a distance usually less than 3500 nautical miles.

Missions include studies of the earth's atmosphere above the limits of balloon flights (about twenty miles) and up to the lowest satellite investigations (about one hundred miles). Sounding rockets also measure the pressure, density, and temperature of the ionosphere, aurorae, and airglow, solar flares, geomagnetic storms, trapped radiation fluctuation, and meteor streams.

With the addition of attitude-stabilization systems some sounding rockets have been used for stellar-astronomy experiments in the X-ray, ultraviolet, and radio regions of the electromagnetic spectrum.

Among the data perhaps best collected by these small relatively inexpensive rockets are those gathered in the recent investigations of noctilucent clouds, which form at about fifty miles of altitude—between balloon and satellite range—and consist of very small ice-covered particles. By analyzing particles collected with a rocket, scientists determined that their origins are extraterrestrial. This was what NASA officials termed "a most unexpected discovery," because radiation pressure from the sun should remove particles of this size, and their presence near earth has not been explained fully.

While the large observatory-type spacecraft are being phased out, NASA's physics and astronomy programs are in a period of transition. A number of important projects normally assigned to unmanned spacecraft were carried out in the Skylab program, and more scientific tasks will be handled in 1980 and beyond by the versatile space shuttle program. All this should greatly decrease the cost of space research.

Meanwhile, fewer missions are being planned. While continuing with the exploitation of data whose complete analysis and implementation into useful outlets will take years, and with the correlative work to support ongoing programs, NASA is developing the technology that will provide a greatly increased capability.

During the 1960s, the physics and astronomy programs made major contributions toward the exploration of earth's cosmic environment. This drastically changed our view of the planet in space and opened our eyes to previously unobserved aspects of the universe.

In the 1970s this program continues with objectives selected along a broad frontier of scientific research. Emphasis is shifting somewhat from the purely exploratory goals of past years toward those objectives that seek either understanding of observed phenomena or increased sensitivity of observation.

Why is space science important? "While we call the work we do space science," says Dr. Naugle, "the scientific research we do in space is no different from the research man has been doing on the earth with steadily increasing capability and determination since he became man. Man has a fundamental drive to explore, understand, and control his environment. This curiosity—man's desire to extend his intellectual horizons—is a measure of the mental health and vigor of an individual or a society."

Dr. Naugle continues, "Out of that fundamental curiosity of man has come all the fundamental knowledge that underlies our modern technology. Science and technology use each other's achievements and progress together. For example, scientists spent almost two hundred years studying and understanding electricity. It took another fifty years or so to develop the technology to put that basic knowledge of electricity to practical use. Scientists in turn used that electrical technology for a variety of purposes, among them the use of large accelerators to study the atom. It took the scientists only fifty years to bring our understanding of the atom to a point where the technology of nuclear energy could be developed in less than a decade. Scientists in turn are now using nuclear technology to produce the tracer elements which have so marvelously advanced our knowledge of disease, genetics, and a variety of life processes.

"It is essential," says Dr. Naugle, "that this nation and the other nations of the world continue to encourage that fundamental curiosity of man so that he will continue to expand his knowledge of the environment, and so the world will continue to have a necessary pool of trained people who will be required in the next decade ahead for solving the problems of food, shelter, communications, and energy for our expanding population, while at the same time maintaining this planet as a habitat for man."

○
Chapter 12
interplanetary ambassadors

The earth is the cradle of the mind, but you cannot live in the cradle forever.

—TSIOLKOVSKY

PIONEERING SOVIET SPACE SCIENTIST

At 7:59 P.M. Greenwich Mean Time, December 14, 1962, an odd-shaped 447-pound robot spacecraft crossed the orbit of Venus at an altitude of 21,645 miles above its surface, following an epic 109-day flight from earth. For thirty-five minutes sensitive instruments aboard the craft scanned the planet, collecting data.

Among the more significant findings were: Venus is blanketed by cold dense clouds in the upper atmosphere; its surface temperature averages 800 degrees Fahrenheit; and, unlike earth, it has neither a measurable magnetic field nor radiation belts.

Such findings were, of course, of great importance to scientists. Never before in the history of civilization had man had access to information on another celestial body attained so close

164

to the source. But far more important was that this lone spacecraft, Mariner II, opened the era of interplanetary travel.

Commenting at the time on the historic Venusian flyby, Dr. William H. Pickering, director of the Jet Propulsion Laboratory in Pasadena, California, which designed and built Mariner and conducted its mission, said: "Soon there will be spacecraft flying by other planets, then orbiting the planets, and finally landing instruments on their surfaces. Exploring the solar system, becoming acquainted with the planets, answering questions about extraterrestrial life—these are the challenges that lie ahead."

His prophecy has proved incredibly precise, for in the years since Mariner II, the United States has dispatched other spacecraft to Venus, Mercury, Mars, Jupiter, and Saturn—all of which assembled and returned to earth mountains of data that are helping scientists completely rewrite astronomy textbooks.

Why is it important to study the solar system—the sun, the planets and their satellites, and the space between them? Astronomers offer an interesting thesis in answer.

They contend that if you consider the solar system from the vantage point of the orbit of Neptune or Pluto, the two outermost planets, you see that it occupies a disc-shaped volume of space 8 billion miles across, with a bright star (the sun) at its center. Circling about this central star at various distances are nine planets and their accompanying satellites, all held in common bondage by the sun's gravitational pull.

No two planets are the same and there are also wide differences among their satellites. Each is at a different stage of development, has a different history, and will have a different future. Considered as a whole, the solar system is a laboratory rich in the number and variety of its specimens.

By studying all of them, one learns more about each one. The moon and Mars give clues to the earth's history, as do the chemical processes taking place on Jupiter. Earth telescopes tell only a part of the story, but knowledge will be increased many thousandfold by spacecraft that photograph the planets from orbit and by those that land and explore the physical characteristics of the atmosphere and the surface.

"We are very fortunate in that all of our planetary spacecraft to date have been successful," says John Naugle. "These missions have substantially advanced our knowledge of the origin and evolution of earth and of our solar system." As an example, Naugle says, "Evidence is growing to support the theory that all the planets had a common origin from the primordial solar nebula, but they condensed from different materials because they formed at varying distances from the hot center of the nebula. For example, the terrestrial planets [Mercury, Venus, Earth, and Mars] being relatively close to the sun and thereby hot, formed from rocky materials with high melting temperatures, such as silicon, iron, etc.

"Farther out from the sun," he says, "the giant planets [Jupiter, Saturn, Uranus, and Neptune] formed from gaseous materials with relatively low freezing temperatures, such as helium, hydrogen, ammonia, methane, etc. Large satellites formed around these giant planets at intermediate temperatures from an apparent mixture of rocky materials and ices."

Scientists agree that to understand the earth better, man must learn all he can about the celestial bodies that are within his technological reach.

For example, problems that complicate earth meteorology, such as the mixture of ocean and continent masses, broken cloud layers, and rapid planet rotation, are isolated and exaggerated on Venus and Mars and are therefore much easier to study. Venus, which rotates very slowly, had no ocean and a continuous thick cloud cover. Mar matches earth's twenty-four-hour day and four seasons, but does not have the complication of the ocean-land masses.

Meteorologists state with certainty that the comparative study of Venus, earth, and Mars will enhance their ability to predict short-term changes in weather, as well as earth's long-term climatic trends.

"Ultimately," says Naugle, "we see the planets serving as safe and convenient laboratories to test theories on the global effects of various proposed manmade measures to control our weather and climate."

"We believe we have just scratched the surface in this field of better understanding earth through the study of our neighboring planets. Our experience to date indicates that these benefits will increase rapidly in the immediate years to come."

Even now Pioneer 10, having passed across the surface of Jupiter in December 1973, is heading toward Saturn. Following in its flight track, Pioneer 11, which left earth in April 1973, has a December 1974 rendezvous with Jupiter. A Viking spacecraft is scheduled to land a package of electronic instruments on Mars in 1976. And in 1977 two advanced Mariner-class spacecraft will be launched to explore Jupiter and Saturn in more detail. They both will fly past Jupiter in 1979 and Saturn in 1981. NASA also is seriously studying combined orbiting and landing-probe missions to Venus in the late 1970s with Pioneer-class spacecraft.

Since its inception, NASA's planetary program has consistently been directed toward three long-range goals, to increase man's understanding of the origin and evolution of the solar system, the origin and evolution of life, and the dynamic processes that affect earth.

In pursuit of these goals, six basic types of planetary missions can be undertaken:

- "flyby" flights in which spacecraft such as the Venus-Mariner II pass close to a planetary target, scan it with sensitive instruments, and relay data to earth;
- atmospheric probe missions, in which spacecraft penetrate a planet's atmosphere and collect data, but are destroyed upon impact with the planet's surface (the U.S. has not launched any such probes to date, although Ranger flights to the moon in the early 1960s, photographing closeup the lunar surface in surveying for future manned landings, were probe-type missions);
- orbiter flights, such as Mariner IX to Mars, in which an instrumented spacecraft surveys the planet from orbit around it;
- lander missions, in which the spacecraft soft-lands on the surface and radios data back to earth; America's unmanned Surveyor moon-landing spacecraft typify this class, as will the Viking to be shot to Mars in 1976;
- unmanned missions that acquire atmospheric and surface samples and bring them back, such as the Soviet sample-gathering flights to the moon;
- manned landing missions—that is, the Apollo flights to the moon.

Only the first four types of missions fall within the limitations of NASA's planetary exploration budget and techical capability during the 1970s.

Of all the planets in the solar system, Venus most closely resembles the earth in size, mass, and surface gravity, but here the similarities end. It is always hidden behind dense opaque layers of clouds that mask it from telescopic eyes—a phenomenon that has frustrated man since Galileo's time.

Counting the Mariner II flight in 1962, six major space probes, three each from the United States and the Soviet Union, have been launched to Venus. The most recent, Mariner X, launched November 3, 1973, sailed within 3600 miles of the planet's surface in February 1974.

Much has been revealed by these probes. From tens of millions of "data words" relayed to earth from investigative spacecraft, it has been learned that Venus's direction of rotation is very slow and opposite that of the other terrestrial planets. Findings from the first man-made object to pass by Venus, Mariner II, also have been confirmed and added to by the five spacecraft that followed its flight path.

Data from radiometer measurements, for instance, correlated with earthbound radar measurements that pierced the cloud blankets, reveal surface relief. This implies that much of the planet's face may be made up of mountains and deserts.

In August 1973 a team of radar astronomers at the Jet Propulsion Laboratory, Pasadena, California, reported that huge shallow craters apparently pock the near-equatorial zone of Venus, a surprising discovery made when high-intensity radar beams were used to pierce the heavy Venusian clouds. "This area of the planet appears to be as crater-infested as the moon," said Dr. Richard A. Goldstein, JPL team leader. He reported the finding by radar of one crater about 100 miles and others varying from 20 to 60 miles in diameter. None appeared to be very deep, the largest being about a quarter-mile. The scientists produced a dramatic map of an area about the size of Alaska which revealed a dozen craters up to 100 miles across.

Most of the surface is extremely hot and dry. It is estimated that at the poles temperatures reach as high as 1000 degrees

Fahrenheit—a point at which lead, tin, and zinc melt—thus it is safe to assume that no organisms known to man could live on the surface of Venus. Additionally, over 90 percent of the Venusian atmosphere is carbon dioxide and only 1 to 7 percent water vapor. This exerts a pressure equivalent to one hundred times that of earth's atmosphere, hence even if man could endure the horribly intense heat on Venus, he would be crushed before he could reach the surface. Such hostile conditions may mean man will never physically explore Venus.

Nevertheless, there is much more to be learned. Despite the successful spacecraft missions to date, many of the major questions concerning this planetary neighbor remain unanswered, and each fragment of information uncovered seems to reveal even more mysteries.

- The appearance of the surface of Venus is still the biggest unknown, perhaps to remain so until spacecraft can be placed in orbit around it. Even then, maps will have to be "drawn" by radar because the surface receives little or no sunlight.
- If the planet is so dry, what makes up the clouds? Obviously they are totally unlike earth's clouds.
- On October 18, 1967, the unmanned Soviet probe Venera 4 destroyed itself like a meteor in the atmosphere of Venus just as it ejected an 850-pound instrumented capsule. Suspended from a parachute, this package drifted down toward the surface for 95 minutes, obtaining pressure, temperature, and density measurements and sampling the atmosphere. It continued to transmit data until it impacted on the surface and then mysteriously fell silent. Was it destroyed by intense heat? Was it crushed by the high atmospheric pressures? Are the answers to these questions, and ultimately keys to the secrets of Venus's atmosphere, forever locked in smashed machinery on a cloud-hidden mountainside?
- From Mariner II and subsequent flights it has been determined by surprised scientists that the magnetic field of Venus is apparently less than one percent that of earth. It is now believed that this small field is a natural consequence of the planet's slow rate of rotation, but more magnetic field measurements are necessary for studying this problem and developing conclusive proof.

Some of the answers to these and other puzzling questions may be soon forthcoming as the enormous harvest of data returned from Mariner X is sifted and analyzed—a process that will take months. Among the yield from this successful mission are thousands of television pictures, the first ever taken of Venus.

More definitive answers undoubtedly will come from flights to Venus of a Pioneer-type spacecraft. NASA has plans to launch two such craft in 1978. One is designed to dispatch four scientific probes, one large and three small ones, toward the surface of Venus and then to enter the atmosphere itself, transmitting additional data to earth until it burns up. The large package will carry up to sixty pounds of scientific instrumentation and will be parachuted down, taking about one and a half hours to descend through the atmosphere. The small probes will fall free to the surface, ending their mission by design when they hit, about seventy-five minutes after entry. These probes are to return data from high atmospheric altitudes as well as the surface.

An instrument-laden sister Pioneer ship is scheduled to go into orbit around the cloud-shrouded planet at about the same time. The primary objective of the twin missions is to gather detailed information on Venus's atmosphere and clouds from which, by comparing it to data on the atmospheres of Mars and earth, scientists hope to be able to construct a better "model" of earth's atmosphere. This could be used in predicting long-term climatic changes as well as the short-term effects of environmental pollution. Experiments to be carried by the spacecraft are to deal with the composition and structure of the Venusian atmosphere down to the planet's surface, the nature and composition of the clouds, the circulation pattern of the atmosphere, and the radiation field in the lower atmosphere.

Surface sampling probes to Venus probably will not be launched until sometime in the 1980s.

Beyond Venus, only 36 million miles from the sun, is the elusive tiny planet Mercury. About a third the size of earth, its surface is solid, rocky, and probably scarred with huge cracks and fissures—the result of the massive impact of comets and asteroids during the early period of the solar system. There is apparently no earthlike atmosphere to shield the planet from such cosmic bombardment.

There is little to recommend the cinderlike Mercury as a

possible future spa for interplanetary travelers, but it is of prime scientific interest. It has a very high density (five and a half times that of water), which suggests a great abundance of such heavy and valuable elements as iron, nickel, tungsten, and chromium. Some scientists believe this planet is about 30 percent rock and 70 percent metals.

Mercury's proximity to the sun may mean that many secrets of the formation of the solar system have been scorched into its surface. Knowledge of the history of variations in solar activity over the past 4 or 5 billion years could well reveal new chapters of information on the climatic history and the evolution of life on earth.

Specific major questions that scientists hope to answer about Mercury through long-distance space probes include:

- What does the planet look like? Is it cratered like the moon, Mars, and earth?
- Mercury's already indistinct features are often suddenly veiled or at least appear to be. Is there an atmosphere that somehow survives despite the sun's incredibly intense heat?
- Since the average density of Mercury is significantly higher than earth's, did it have a different origin or were its lighter elements volatilized by the sun, perhaps by a much hotter sun?
- What is the temperature of Mercury's surface, and how does it vary from local night to day?
- How strong is Mercury's magnetic field, and what is the nature of its interaction with the solar wind so close to the sun?

It was hoped that many of these and other questions would be answered in March 1974, when Mariner X, following its flyby of Venus in February, passed within 416 miles of Mercury's surface. Television cameras took hundreds of photos, the first ever of this planet.

These vivid photographs, plus data from other measurements, led scientists to conclude that Mercury looks like the moon, but has internal characteristics reminiscent of the earth. In the pictures the rugged craters of Mercury showed themselves in sharp contrast to the Maria or smooth regions, which scientists believe probably formed sometime after the earliest stage of the planet. From these pictures and other data, NASA expects to improve maps of Mercury from the vague image currently available from earth-based telescopes, to a resolution of approximately 300 yards. Mariner X was also assigned the tasks of determining the temperature of Mercury's surface and its mass, diameter, and density to detect the presence of a magnetic field, and measuring the planet's environmental interaction with the solar wind.

Aside from the virgin scientific data Mariner X collected from its mission to Mercury, it also was to demonstrate two significant engineering advances. It was to be the first spacecraft to use a planet's (Venus's) gravitational field to speed it up and propel it—in a "crack-the-whip" fashion—toward another celestial body.

Second, following its flyby of Mercury, Mariner X was to sail on a path that would carry it around the sun, and then in late September 1974 return for a second Mercury encounter six months after the first.

Beyond Mercury, of course, is the sun—the provider of most of the earth's energy, controller of its weather, disturber of its communications. The more we know about the sun and how it influences the earth, the more we can predict events on earth.

While a few spacecraft have ventured in the general area of the sun in past years, none has come as close as will a Helios spacecraft to be launched late in 1974. It is a joint venture of the United States and the Federal Republic of Germany: the United States will launch the Helios, which is being built by Germany.

The spacecraft is scheduled to move inside the orbit of Mercury and approach within 23 million miles of the sun. A second Helios launch is planned in 1975.

○
Chapter 13

the search for other life

I cannot say I believe that there is life out there. All I can say is that there are a number of reasons to think it is possible and that we have at our command the means of finding out. . . . I would be very ashamed of my civilization if we did not try to find out.

—ASTRONOMER CARL SAGAN

In classic Roman mythological order Mars ranked second only to Jupiter as the most important deity. The god's planetary namesake, however, fourth of nine major bodies stretching from the sun, has no peer in terms of the spellbinding fascination it historically has held for earthlings. No other celestial neighbor, with the possible exception of the moon, has generated such interest, triggered such controversy or been studied with such fevered intensity as the mysterious Red Planet.

In the earliest days of telescopic observation, nearly four centuries ago, Galileo was intrigued with Mars. Its unusual marked surface and thin atmosphere permitted scientists to measure its twenty-four-and-a-half-hour day and annual change of seasons as early as 1659. Sir William Herschel, a hundred and

Dramatic portrait of Mars was taken from Mariner VII spacecraft at a distance of 337,132 miles from the planet's surface.

fourteen years later, detected seasonal variation in the sizes of the Martian polar caps.

Dark areas of the planet were regarded by many pioneer observers as expanses of water. In 1877 G. V. Schiaparelli, an acknowledged expert, designated linelike markings on Mars's surface as canals. "Their singular aspect," he wrote, "and their being drawn with absolute goemetrical precision, as if they were the work of rule of compass, has led some to see in them the work of intelligent beings—inhabitants of the planet."

The theory of life on Mars was so popular in the early 1900s that the idea was once fostered to attempt to "communicate" with beings there by digging a canal in the shape of a huge right triangle in the Sahara Desert. In fiction the planet has for centuries been populated, irrigated, and civilized. It has long been thought of as home base for flying saucers and, as recently as 1939, radio listeners on earth were persuaded that they could be, and indeed had been, invaded by Martians.

Through today's highest-powered telescopes Mars appears as a ruddy or orange-colored disc relieved by darker markings of greenish-blue. However, such long-distance views—in its elliptical swing around the sun, Mars never comes closer than 35 million miles to earth—are distorted by thick veils of atmospheric haze.

So, until the dawning of the Space Age the best astronomers could do was speculate about the true, detailed characteristics of the planet. They knew Mars has about half of the earth's diameter and only about a ninth of its mass, but little more could be agreed upon. All this was dramatically changed on July 14, 1965, when a 600-pound electronic package named Mariner IV, launched from Cape Kennedy seven and a half months earlier, swept within 6118 miles of the Martian surface. Twenty-one of what experts called "the most remarkable scientific photographs of this age" were taken, and invaluable data on the planet's magnetic field, strength, radiation, and cosmic dust were collected.

Perhaps the most surprising phenomenon of the mission was the discovery of dense-packed lunarlike impact craters on the craggy face of Mars. More than seventy craters appeared in the pictures, which covered less than one percent of the planet's surface. No clear evidence of the famous canals was apparent and no physical features were found that could have been the basins of former oceans or the beds of ancient rivers, lakes, or seas.

Radio beams directed from Mariner found very thin air, an ionosphere and atmosphere considerably less dense than expected. Comparison of various atmospheric properties suggested that carbon dioxide was the major constituent. The spacecraft's instruments searched in vain for evidence of earthlike radiation belts or magnetic fields. The findings from Mariner IV, though not conclusive, indicated the planet to be unsuitable for any known major life form as we know it.

Five years later, in July and August 1969, the twin interplanetary voyagers, Mariners VI and VII, flew even closer to Mars, giving added dimension to the growing bank of planetary knowledge. They took over 2400 measurements on the composition of the Martian atmosphere, and 800 closeup temperature readings. TV cameras snapped and transmitted 198 high-quality pictures of the surface. Among the findings:

- The probability that microbial life exists is much less than had been previously supposed. It now appears unlikely that Mars has ever had an ocean, for instance. The amount of water vapor detected in the atmosphere is too little to permit the growth of any known terrestrial species. It was also determined that lethal ultraviolet radiation reaches the planet surface for its atmosphere is too thin to protect it.
- Exciting new features were discovered, such as the chaotic and the featureless terrains, which constitute dramatic evidence of presently unknown surface phenomena.
- Pictures taken at the edge of the planet showed an atmospheric haze almost six miles thick.
- Surface temperatures were found to be relatively moderate—between −63° and +77° F. during the day, and ranging from −63° to −153° F. at night.
- Numerous craters up to 300 miles in diameter were seen. The heavily cratered polar-cap region appeared to be covered by a thin layer of ice.

Despite this impressive dossier, more questions have been raised than answered by the probes of the 1960s. The flyby missions at best recorded data from only a tiny segment of Mars and for only a few hours.

How fair is it to draw conclusions from such a limited sampling? What, for example, if a spaceship from another world flew by earth, passing only over the Sahara Desert? Could it be that we were getting a distorted picture of Mars, one not truly representative of the planet's features?

These questions were emphatically answered by data collected and sent to earth from the flight of Mariner IX. A trim blue-winged spacecraft that carried only 150 pounds of sophisticated instrumentation, it was launched from Cape Kennedy May 30, 1971, when earth and Mars were within 35 million miles of each other, the closest they had been since 1924.

Following a 167-day flight through space, a special rocket

engine was fired, slowing Mariner IX's speed, placing it in orbit around Mars on November 13—the first man-made object ever to orbit another planet.

The first photographs it transmitted to earth revealed that the spacecraft received a hostile greeting. The entire Martian globe was being ravaged by a dust storm—the longest, most intense and widespread storm in the history of man's observations of the planet. So dense, in fact, was the great yellowish pall that enveloped Mars that from earth only five surface features could be distinguished—the south polar cap and four dark spots, which proved to be huge volcanoes.

But even instrumented data on this storm proved of immense value. The dust clouds swirled to altitudes of 30 to 40 miles, cooling the surface of the planet and warming the atmosphere. Measurements of such phenomena, scientists believe, can help them better understand and calculate the effect of increasing pollution on the earth's global climate.

The massive storm raged for several weeks before subsiding; then, early in 1972, the veil lifted, and Mariner IX began recording an incredible picture of a Mars man had never before seen or imagined. Based on the limited evidence that had been collected during the earlier Mariner flybys in 1965 and 1969, the planet prematurely had been generally considered lifeless with a monotonous terrain devoid of mountain ranges, great faults, and volcanic activity. It was believed that no life forms could exist there and that it was geologically dead, much like the moon.

Mariner IX, to the astonishment of nearly everyone, proved just the opposite to be true.

This most prolific of all planetary spacecraft circled Mars twice a day for almost a year, radioing to earth a continuous stream of scientific data—including more than 7300 spectacular pictures—that have revised all previous concepts of Mars.

Instead of being dead, it has been found to be a geologically active planet internally alive, with a surface that is constantly being altered by dynamic forces. It appears to be more like the earth than the moon, yet different from both.

It has volcanic mountains that dwarf anything on earth. One giant volcano, Nix Olympica, towers over 70,000 feet from its base and is thus nearly two and a half times higher than Mt.

Everest. Its lava flow field extends roughly in a circle over 300 miles across.

A great rift canyon, measured by Mariner IX, is three to four times deeper than the Grand Canyon and more than 3000 miles long. Tributaries branching from the walls of the canyon look very much like water-formed features on earth.

The discovery of these "riverbeds" was perhaps the biggest surprise, and scientists say close examination has shown they are not faults in the planet's crust. It is now widely believed that they are exactly what they appear to be in the closeup photographs: the beds of rivers over which water once flowed. How long ago the rivers ran dry and how much water may be found in the Martian core and thin atmosphere are questions still to be answered.

But Mariner IX solved a lot of long-standing puzzlers. Instead of the dreary cratered surface pictured by the earlier spacecraft, which had swept by about one percent of the planet's terrain, Mars actually can be subdivided into at least four major geological provinces: volcanic, an equatorial plateau region with faults and rifts, the featureless cratered and smooth terrains seen by Mariners IV, VI, and VII, which has been found in northern and southern hemispheres and may be more ancient than the other areas, and the south polar area, which is also cratered and is blanketed by glacial sediment layers (similar deposits appear in the north polar regions).

For centuries man was perplexed by what appeared to be continuing changes in the Martian surface. Many thought they were the result of the growth and spread of vegetation on Mars during spring and summer seasons. But we now know with near certainty that these "variable markings" as observed from earth are the result of the planet's frequent raging dust storms and cloudiness.

Mars is a far more dusty place than scientists first realized. Infrared observations from Mariner IX show that large areas of the globe are coated with a very fine dust, and craters are covered with vast dunes.

Desolate, scarred face of Mars was photographed by Mariner IX spacecraft from orbit around the planet.

Mariner IX also photographed Mars's two tiny moons, Phobos and Deimos, for the first time. Like earth's moon, they were found to be synchronous with the planet—that is, one side always faces the planet. Both are heavily cratered, apparently from meteorite impacts, and have irregular rocklike bodies, which suggests to scientists that they are captured asteroids rather than bodies formed from the same material that coalesced to form Mars.

An ultraviolet spectrometer experiment aboard the spacecraft made 30,000 individual measurements of relative surface brightness, producing a preliminary topographical map of Mars.

An infrared interferometer spectrometer experiment gathered data on Martian temperatures from which we learned that the planet's north pole is much colder—about −200 degrees Fahrenheit—and drier than the coldest spot on earth, Antarctica (−125 degrees Fahrenheit).

Temperatures on the surface of Mars, recorded by an infrared radiometer, ranged from +81 degrees Fahrenheit, in the equatorial zone to −189 degrees Fahrenheit at both poles. These readings did not vary substantially from those recorded by instruments on Mariners VI and VII.

On October 27, 1972, after 349 days of operation, Mariner IX's engineering telemetry signals ceased. At the time the spacecraft was 238,416,000 miles from earth. Engineers on earth sent the final command, the last of 45,960 commands issued, to turn off Mariner's radio transmitter—thus ending one of the most spectacularly successful space missions ever undertaken. The spacecraft is expected to remain in Mars orbit, mute, for at least fifty years.

"Mariner IX has rewritten the textbook on Mars, and has shown it to be a much more dynamic, changing planet than it was previously believed to be," said NASA's John Naugle.

But for all the definitive data that Mariner IX returned to earth, for all the myths it shattered, it did not provide a conclusive answer to the most intriguing question of all: Is there life on Mars?

Is earth truly a unique life-supporting planet in the immense totality of creation? There is growing evidence to the contrary.

Our galaxy contains 100 billion stars, many of which are surrounded by families of planets, according to the best astronomical evidence. In studying these stars with telescopes, man has been able to verify that the basic chemicals of which earth is composed are found throughout the universe. In the last century it was proved that the ratio of these elements in our own solar system is consistent with the overall ratio generally observed throughout the universe.

The existence of complex organic material has been positively identified in the vast regions of interstellar space. These organics have been detected in meteorites, thereby confirming the widespread presence of organic material in our solar system as well as in the universe at large.

Further, radio astronomers have detected simple organic compounds in intersellar space. Recent detection of complex organic compounds has increased scientific confidence that life could evolve on other worlds. But science cannot calculate the probability of encountering extraterrestrial life in this or other solar systems on the basis of this evidence. Man cannot tell conclusively by laboratory studies or theoretical reasoning whether the evolution of life is vanishingly improbable or quite likely. He can estimate the probability only by searching for signs of extraterrestrial life.

The nearest reasonable planet on which to look is Mars, which is dry, cold, and less favorable than earth for the support of life, but not implacably hostile. Life could exist in the harsh climate of Mars, and if it does, man will know that on planets with comfortable climates—similar to that on earth—the chances of finding life are substantial. He will have strong reason to believe that many inhabited solar systems, perhaps billions, lie around him in the galaxy.

But even if no life of any form is found on Mars, exploration may settle a question of equal importance for determining the probability of life arising out of nonliving chemicals. Is it possible that Mars, lifeless today, was once the site of a rich variety of life that disappeared later in its history?

This question is keyed to the abundance of water. Mars is relatively dry today, but discoveries by Mariner IX of volcanism and riverbeds on the planet's surface suggest that it could have had a substantial supply of water that at times became available in liquid form. The water could have remained long enough to permit some form of organism to evolve, only to be snuffed out later when the vital gases and water on Mars disappeared. If that happened, traces of life on the surface may still be found.

Even if no signs of life, extant or extinct, are found, it is crucially important to study the nature of other planets presumed to have originated at about the same time and by the same processes as earth. In this context, finding no life forms on Mars could be nearly as important as the discovery of life. The study of a planet not too dissimilar from earth which has evolved in the absence of life would provide scientists with a yardstick with which to determine, for example, how the atmosphere of earth

Scale model of NASA's Viking spacecraft, scheduled to orbit the planet Mars in 1976.

has been influenced by the presence of biological processes. Comparative planetology will be of great value in understanding earth and in formulating measures to protect its environment.

Based on current knowledge, however, scientists believe it highly probable that some evidence of complex organic compounds will be found on the surface of Mars.

If these do in fact exist, they will be found in 1976 by the mechanical limbs of the two spacecraft named Viking. In one of the most ambitious and complex missions ever attempted, these instrument-carrying robots will be launched into Martian orbit, as Mariner IX was. Then each spacecraft will split, leaving an orbiter to circle the planet while a lander descends to the surface to spoon up and analyze soil samples.

Each Viking, built by the Martin-Marietta Corporation of Denver, will weigh 7500 pounds and will be launched toward Mars atop a Titan III-Centaur rocket. Each will travel 440 million miles through space, taking almost a year to reach its destination. Once their Mars orbits have been stabilized, all instruments checked out, and the landing sites selected, the landers will be dispatched from the orbiter vehicles for the drop to the surface.

Their descent and landing will not be unlike the astronauts' lunar module landings in the Apollo program. A specially designed aeroshell will shield the lander against the intense heat generated by its high-speed deceleration through the thin atmosphere, and at 20,000 feet a parachute will be deployed. Then, a mile above terrain, small rocket engines will fire, slowing the landers—each weighing 147 pounds—to allow a soft touchdown. These engines will automatically shut off as the vehicles' sensitive footpads touch the surface.

Throughout the mission the orbiters will continue to sweep around the planet, each laden with two high-resolution television cameras and other sensitive equipment. They will collect data on surface temperature, atmospheric water concentration, the presence and movement of clouds and dust storms, the topography and color of the terrain, and other information.

But the landers will command prime scientific interest. Essentially, the landers' mechanical arms, guided by mission directors millions of miles away on earth, will feel their way

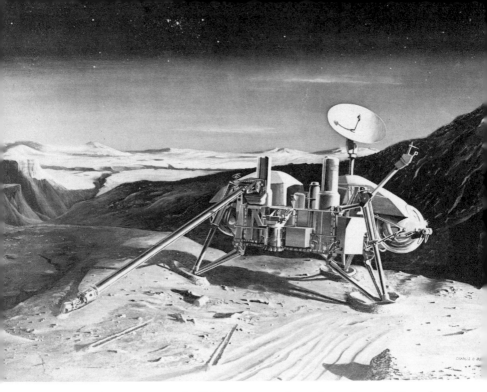

Artist's conception of a Viking spacecraft on the surface of Mars shows how the spacecraft's scoop-like "arm" will examine the Martian terrain.

about the surface in the vicinity of the spacecraft, digging up scoopfuls of Martian terrain which will be thoroughly analyzed by other instruments. NASA officials say the combined total of these instruments represents an analytical capability that normally requires several large and very well-equipped earth-based laboratories.

Actually, in what has been described as one of the most complex pieces of electromechanical-chemical machinery ever devised—the equivalent of three automated chemical laboratories compressed into one cubic foot—samples will go into a rotating conveyor that distributes measured portions to several test cells. Specifically, lander instruments consist of a gas chromatograph/mass spectrometer for detecting and identifying organic molecules—the building blocks of life—in the soil, a biology instrument capable of performing three different life-detection experiments, three meteorological sensors, a seismometer, an X-ray fluorescence spectrometer for inorganic chemical analysis of surface material, two facsimile cameras, and a magnetic collector head on a boom for gathering soil samples and measuring surface properties.

The search for life will be carried out through a complicated series of experiments. A photosynthetic analysis will determine if organisms exist on Mars through nourishment from organic materials.

The molecular structure of the soil and atmosphere will be analyzed through a gas chromatograph/mass spectrometer, which will "sniff" the atmosphere and also analyze gases given off by soil samples collected by the scoop.

Daily measurements of temperatures, pressures, humidity, and wind velocity and direction will be taken. Physical and seismic characteristics will be analyzed through the reaction of the boom-mounted surface sampler as it digs into the soil. A miniature seismometer mounted in the latter will detect ground motion transmitted through the legs of the spacecraft.

Each lander's two cameras, substituting for man's eyes, can be directed to focus on a specific area closeup or perform panoramic scanning sweeps of the entire landscape around the spacecraft. The pictures will convey a great deal of information about the geological character of the face of Mars, and they could aid in identifying any higher form of life that may exist.

The landers will also be the cleanest spacecraft ever launched. Each will be heat-sterilized before hand, in compliance with international planetary quarantine requirements, under which all earth-launched spacecraft are subjected to biological surveys and monitoring control. Not only does this safeguard other planets against the possibility of contamination by terrestrial germs, it also prevents inclusion of false data in the life-detection experiments aboard Viking. In fact, under international agreements, the United States has pledged not to land an unsterilized spacecraft on Mars until at least 2018.

Whatever Viking finds—evidence of life, or no traces of life at all—will not diminish the enthusiasm for or the inevitability of a manned landing on Mars. Rather, the findings will undoubtedly only whet man's interest.

Exactly when such a mission will take place has neither been determined nor is likely to be for some time. In July 1969, in the flush of the spectacularly successful first manned lunar landing by Apollo 11, the administration said a manned landing on Mars should be one of America's next major goals in space. There was talk of a flight that would be launched in 1986, when the Red Planet will be in a highly favorable position relative to earth. But in light of other national priorities, it appears that such a mission will not be launched before 1990, and perhaps not even in this century.

Despite the colossal nature of the technological challenges, they are not insurmountable. NASA experts believe they could send man to Mars and get him home safely if given a ten-to-fifteen-year lead time. It must be remembered that only recently a manned mission to the moon was virtually inconceivable, yet it was only eight years from the time it was declared a national scientific objective and its resources and funding were committed, to the time the goal was reached.

And so, ultimately, neither engineering problems nor technology breakthroughs, but, rather, politics will determine when man will explore Mars. The Mars mission can be accomplished only when the nation's leadership assigns it the priority status necessary for completion, as was true during the lunar landing program. It will require billions of dollars and years of work, and it will require uncommon courage and foresight to commit such resources.

Whether it happens in this century or not, it will happen.

Chapter 14

reaching the outer limits

Beyond Mars and orbiting the sun once every twelve years from a distance of 480 million miles is colorful, dynamic, provocative Jupiter, the colossus of the solar system. It has more than 11 times the diameter of earth, 318 times the mass, and 1000 times the volume. It could swallow up 1300 earths. In fact, it has more than twice the mass of all the other planets combined—so massive that it is almost a small star.

Like Mars, it has intrigued and excited man since its discovery. Galileo made the first telescopic observations of Jupiter and discovered its four larger moons in 1610. Two of these, Ganymede and Callisto, are about the size of Mercury, while two others, Io and Europa, are similar in size to earth's moon.

From hundreds of years of astronomical observations and

Clues to the origin of the solar system, many scientists believe, may be found by spacecraft sent near the mysterious planet Jupiter.

analyses man has learned much about Jupiter. It rotates completely once every ten hours—the shortest day of any of the nine planets—and at so great a speed that the planet bulges at the equator even though its diameter is eleven times earth's, but it has a total of twelve moons. Its visible surface covers about 24 billion square miles.

Jupiter is vastly different from any of the rocky terrestrial planets. Most scientists agree that is is comprised of a mixture of elements similar to those in the sun or in the primordial gas cloud that formed the sun and planets. Hydrogen and helium—light gases that are the main constituents of the universe—make up at least three-quarters of Jupiter's chemical composition.

Through spectroscopic studies of Jupiter's clouds from earth, deuterium, methane, and ammonia have been identified. These basically are the same elements believed to have produced life on earth over 4 billion years ago. Jupiter appears to have its own internal heat source. It apparently radiates about three to four times as much thermal energy as it receives from the sun, thus leading experts to believe that large regions below the frigid cloud layers hovering over the planet are actually in temperature ranges conducive to the production of living organisms. There may be some forms of life there.

Jupiter and its giant planetary neighbor, Saturn, are viewed as huge chemical factories that are continuously generating organic compounds, which might be the first steps toward the evolution of life.

Conversely, the planet may have no solid surface. Because of its high specific gravity, the atmosphere can change from a thick gaseous atmosphere to oceans of liquid hydrogen, then to a slushy layer, and finally to a solid hydrogen core. Ideas of how deep beneath its striped cloud layers any solid hydrogen "icebergs" or "continents" might lie vary by thousands of miles, although the general range is from 60 to 3600 miles. No one knows for sure at what depth beneath the clouds Jupiter's enormous pressures have turned its hydrogen to a metallic solid.

Because of the speed of its rotation, Jupiter is striped or banded, parallel with its equator, with large dusky gray regions at both poles. Between the two polar regions are five bright salmon-colored stripes, known as zones, four darker slate-gray stripes, known as belts. With its vivid coloring and its dense, cloudy ambulent atmosphere, Jupiter appears to be floating in space like a great multihued rubber ball.

The planet's most bizarre feature is the Great Red Spot, often called the "Eye of Jupiter." This enormous woundlike oval, 30,000 miles long and 8000 miles wide, has for centuries spurred a wide range of scientific speculation and controversy. The fact that the Great Red Spot periodically disappears only to resurface at a different location adds to the intrigue. It might be an enormous vertical column of gas or a hydrogen ice "raft" floating on a warm hydrogen bubble in the cooler hydrogen atmosphere. Its distinctive color might result from the presence of organic compounds manufactured in the Jovian atmosphere.

The precise nature of the Red Spot is unknown, but it is not likely that such a feature, several times the size of earth, could be so well defined in rapidly rotating clouds unless it were generated by a surface phenomenon.

If Jupiter radiates more energy than it receives from the sun, what is its source? What really lies beneath Jupiter's thick gaseous atmosphere? Are there, as many believe, oceans of liquid hydrogen? Is there a solid hydrogen core? How does Jupiter broadcast the predictably modulated tremendously powerful radio signals? Are these radio emissions involved with the planet's powerful magnetic field and radiation belts? Why are the radio waves pulsed every time the satellite Io passes around the planet? Why, when tiny shadows are cast by one or more of Jupiter's twelve satellites, do telescope measurements indicate the temperature in the shadow area to be *warmer* than in the surrounding lighted clouds? Why is Jupiter's color distinctly orange and much redder than any other object in the solar system? Is there water below the atmosphere? Is Jupiter a living fossil of the solar system—or does it contain life? These are only a few of the basic questions to which scientists are seeking answers.

The first giant step toward obtaining such answers—the first step toward direct exploration of the outer solar system—was taken on the night of March 2, 1972, with the launching of an Atlas-Centaur at Cape Kennedy. Aboard the rocket was Pioneer 10, a spindly 570-pound spacecraft carrying a 65-pound load of sophisticated scientific instruments designed to make twenty different measurements of Jupiter's atmosphere, radiation belts, heat balance, magnetic field, moons, and other phenomena. The payload was to determine the character of the heliosphere (solar atmosphere) and to investigate the interstellar gas, cosmic rays, asteroids, and meteoroids between earth and Jupiter.

Studies of Jupiter's rapidly rotating atmosphere could lead

Long-shot: Pioneer 10 spacecraft, with a mission to fly by the planet Jupiter after a 21-month journey through space, was launched from Cape Kennedy aboard an Atlas-Centaur rocket March 2, 1972.

to a better understanding of earth's weather cycles and to insights into earth's atmospheric circulation. Findings could also lead to such Jovian resources as a quantity of petrochemicals equivalent to earth's consumption for a million years.

The Atlas-Centaur drove Pioneer 10 away from Earth at 32,000 miles per hour—faster than any man-made object has ever flown. It passed the moon's orbit in eleven hours, and for the first week of its spectacular flight it averaged a half-million miles a day.

Less than three months after launch Pioneer 10 streaked past the orbital path of Mars, eclipsing all previous distance marks set by American and Russian spacecraft. About four and a half months after launch and nearly 120 million miles from earth, Pioneer 10 entered the mysterious Asteroid Belt, to the consternation of scientists on earth. The belt, an immense ring of stony rubble circling the sun between Mars and Jupiter, constitutes a special region of rocklike fragments and particles roughly 152 million miles wide. Scientists believe that asteroids either condensed individually from the primordial gas cloud that formed the sun and planets; or, somewhat less likely, that they are debris from the destruction of a very small planet. Clearly they contain important information on the origin of the solar system. Although the largest of these bodies, Ceres, has a diameter of 480 miles, the majority are merely boulders, rocks, and pebbles. But there may be as many as fifty thousand with diameters of at least one mile—and a collision with even one would have rendered Pioneer 10 useless.

But the spacecraft survived this potentially perilous portion of its journey. During its seven-month 205-million-mile ride through the belt, it received no damaging hits from high-velocity asteroid particles. En route the Pioneer 10 collected data on the features of the solar atmosphere, found various elements and isotopes among solar particles, and gathered information about the "interstellar wind" outside the heliosphere. It also measured the particle distribution in the Asteroid Belt.

In December 1973, following an epic 620-million-mile, 21-

Artist's conception of Pioneer spacecraft sailing above Jupiter's intriguing Red Spot.

month flight, Pioneer 10 swept within 87,000 miles of Jupiter. During the four days of the flyby, its instruments recorded and transmitted thousands of bits of data back to earth. That in itself was an accomplishment of unprecedented proportion. For radio signals from mission controllers, traveling at the speed of light, take forty-five minutes to flash from earth to Jupiter—90 minutes round trip. The instructions were received by a nine-foot dish antenna protruding from Pioneer 10. Return signals from a half-billion miles out in space were "being listened for" by the great dish antennas of NASA's deep space network.

The spacecraft's instruments found that Jupiter's dark orange-brown belts are "hot"; that spaceships from earth may be able to pass through Jupiter's intense radiation belts at certain points close to the planet; plus a number of other surprising things. As with the enormous amount of raw data collected on the moon and on previous flights to Mars, scientists will be analyzing Pioneer 10's findings for months, even years. It has given us a wealth of new knowledge about Jupiter and many aspects of the outer solar system and our galaxy. It has returned the first close-up images of Jupiter and made the first measurements of Jupiter's twilight side, never seen from the earth.

Pioneer 10 also opened the era of exploration of the outer planets—even now it is en route toward the ringed giant, Saturn —and following its flyby of Jupiter, it was to collect invaluable data on galactic cosmic rays, solar winds, and the distribution of interstellar neutral hydrogen and helium.

Sometime in 1977 Pioneer 10 may cross the orbit of Saturn, at a point 930 million miles from the sun, but most of its instruments will no longer be operative. In another three and a half years it will reach the orbit of the planet Uranus, 1.8 billion miles from the sun. The spacecraft will become a deaf-mute—simply because it's too far out to reach—and all communications with earth will cease.

Sometime in the 1980s Pioneer 10 will shoot past Pluto and skip off into the eternity of space beyond our solar system. Once its nuclear power supply of energy has depleted, it will drift forever toward the stars.

Attached to the wandering spacecraft is a gold-anodized aluminum plaque, depicting the nude figures of a man and woman of earth. The man's hand is raised in a gesture of friendliness, and the sun and the nine planets of our solar system are shown too. This pictorial message is intended for any other intelligent species that might exist, who might find Pioneer 10, thousands of years from now in some other star system.

On April 5, 1973, NASA launched a second spacecraft—Pioneer 11—to Jupiter. It is expected to rendezvous with the planet in December 1974, and among the scientific instruments aboard is one that returns pictures of the planet. Other equipment will measure the planet's atmospheric characteristics, solar and galactic cosmic-ray particles, interplanetary and Jovian magnetic fields, the solar wind, and interplanetary asteroidal material, Jupiter's heat balance and the interstellar wind. Pioneer 11's final flight course—just how close to the planet's surface it will fly—depends upon final evaluation of data collected by Pioneer 10, which could be programmed to sweep by with a periapsis (closest approach) of 22,000 miles. The key question is, How close can the spacecraft travel to the powerful Jovian radiation belts—believed to be as much as a million times stronger than earth's Van Allen radiation belts—without being crushed?

In this decade the United States plans two more missions to the far ranges of the solar system. Two Mariner-class vehicles are to be launched to Jupiter and Saturn in 1977. According to the preliminary flight plans, the two Mariners will be launched about two months apart and will take about a year and a half to reach Jupiter. Then, because of the acceleration provided by Jupiter's gravity and orbital velocity, the spacecraft will literally be flung, slingshot fashion, toward Saturn. They will encounter the ringed planet about four years after being launched from earth.

Although the two ships, in case one mission should abort, will have identical instrumentation, they may have differing trajectories, allowing each to pursue a different set of specific scientific objectives. Spacecraft and experiment designers will capitalize on the findings of Pioneers 10 and 11.

Astronomical interest in Saturn, with its lovely translucent rings, has always been high, but recent scientific disclosures regarding the planet have greatly increased the thirst for more knowledge about it. Astronomers say Saturn has at least three rings ranging outward from the planet for 85,000 miles. The width of the principal inner ring is estimated at 16,000 miles, and the outer ring at 10,000 miles. Despite their globe-girdling circumference, the rings appear through telescopes to be perhaps only a half mile thick.

Many experts have long believed Saturn's rings to be very thin and to consist of ice crystals, dust particles or gas—or some combination of these. But in 1973, following the first successful radar probing, it was found that the rings appear to be made of solid chunks, perhaps rough and rocky.

Using NASA's 210-foot antenna at Goldstone Station on California's Mojave Desert, two radar astronomers, Dr. Richard M. Goldstein and George A. Morris, Jr., directed the 400-kilowatt radar beams at Saturn when it was 700 million miles from earth. They reported receiving much stronger bounceback signals than expected from such a distance. And they concluded, from the radar results, that the rings cannot be made up of tiny ice crystals, dust or gas. The radar echoes indicate rough, jagged surfaces with solid material a yard in diameter or larger, possibly much larger.

Dr. Goldstein warned, moreover, that Saturn's rings must be considered an extreme hazard to any spacecraft sent into or near them. The two Mariners also will fly close-in past satellites of Jupiter and Saturn and this too has spurred great scientific interest.

Although earth's moon is dry and lifeless, some experts reported new evidence that some outer planetary satellites are covered with water ice and may have thin atmospheres. At least one—Titan, the largest of Saturn's ten moons—might support lower forms of life. In January 1973 Cornell University's astronomers said they believe Titan's atmospheric conditions to be similar to those of earth at the dawn of life.

The trajectory planned for the Mariner missions could take the spacecraft within one hundred miles of Titan. It is possible that instruments could detect life forms, if such exist, during such a pass-by, sometime in 1981.

No spacecraft missions to the outer planets are planned beyond the 1977 Mariner flights. Inevitably, however, man will probe farther. Depending on the Pioneer and Mariner findings in the 1970s, it is reasonable to assume that spaceships will be sent to orbit Jupiter and Saturn later in this century, and that these will in turn be followed by unmanned landers. Other ships will be dispatched to the far reaches of the solar system to unveil characteristics of Uranus, Neptune, and Pluto—planets about which we know virtually nothing. It will be decades, perhaps even a century or more, before this happens. But man will follow the interplanetary trail blazed by robot spacecraft, using the planets as safe harbors as he reaches beyond the solar system into the universe.

Part Four

the
one world

O
Chapter 15

space international

Imagine looking back on a colorful sphere silhouetted against the black vastness of space. You see no boundaries between nations . . . no races . . . no different religions . . . no opposing political philosophies.

—ASTRONAUT JACK SWIGERT
CIRCLING THE MOON

From sweeping orbital platforms a few hundred miles up, sensitive electronic eyes ceaselessly scan earth's beautiful multihued surface—tracking the uncertain movement of a baby tropical storm northeast of Venezuela; searching across the vast, uninhabited deserts of Saudi Arabia for rich new mineral sources; probing the depths of fresh snows in the remote Himalayas.

Live coverage, not only of major news events but of championship sporting events—soccer in South America, Grand Prix races in Europe, tennis matches in Australia—is beamed to tens of millions of people in scores of nations around the globe.

Using powerful microscopes in modern laboratories in Oslo, Taiwan, Johannesburg, Prague, and Seoul, eminent scientists analyze moon rock fragments brought back by American astronauts.

In Argentina, Japan, Italy, Canada, and many other nations, skilled technicians prepare a great variety of scientifically important experiment-payloads for rocket launchings at sites from Cape Kennedy to Pakistan to the Brazilian coast near Natál. These examples are representative of hundreds of continuing and planned programs heralding the fact that there are no national boundaries in space. Despite American and Russian technological superiority today the infinite domain of man's last and greatest frontier is truly open to everyone.

"The United States will take positive, concrete steps toward internationalizing man's epic venture into space—an adventure that belongs not to one nation but to all mankind," President Nixon said, continuing a policy previously endorsed by Presidents Eisenhower, Kennedy, and Johnson. "I believe both the adventures and the applications of space missions should be shared by all peoples. Our progress will be faster and our accomplishments greater if nations will join together in this effort, both in contributing the resources and in enjoying the benefits."

The United States has fostered such objectives since the 1958 creation of the Space Act, which states in part, "The aeronautical and space activities of the United States shall be conducted so as to contribute materially to . . . cooperation by the U.S. with other nations and groups of nations in work done pursuant to this act and in the peaceful application of the results thereof."

In fact, NASA has entered into more than 250 agreements for international space projects. It has orbited foreign satellites with American rockets, flown foreign experiments on American spacecraft, and participated in more than 700 cooperative scientific rocket soundings from locations in all quarters of the world.

The United States also openly shares with nations large and small, rich and poor, its enormous wealth of space knowledge and of expertise for practical benefits. In orbit now, for example, is the first Earth Resources Technology Satellite, ERTS-1, which is returning data requested by three hundred experimenters representing the United States, thirty-seven foreign countries, and two United Nations groups.

Instruments aboard this pioneering spacecraft have concentrated on geological formations that may contain mineral deposits, inventories of agricultural crops and forestry reserves, surveys of marine resources and life, observations of environmental and ecological conditions, and the instantaneous detection of such natural disasters as forest fires, floods, earthquakes, volcanic activity, and plant diseases and soil erosion.

Most foreign experimenters are concerned with global geology, but more than sixty participating scientific investigators are directly involved with ecology and the quality of the environment.

The experimental ERTS-1 makes fourteen revolutions of the earth a day, thus providing 80-percent global coverage, and may be the forerunner of an operational system that could give man an invaluable tool for properly managing, for the first time in history, his planet's resources. Data are made available on request to all participating nations.

Fully recognizing the enormous potential of this program, the United Nations Committee on the Exploration and Peaceful Uses of Outer Space has encouraged rapid progress in the surveying of the earth's resources, especially among developing nations. The United States works closely with the Outer Space committee, established in 1961 to "provide for the exchange of information on space activities, to study measures for the promotion of international cooperation, and to report on the legal problems that might arise from the exploration and use of outer space."

Globe-circling United States meteorological satellites continually flash streams of critical weather information to ground stations. Through advanced instrumentation, developed during more than a decade of experimental missions, these versatile spacecraft routinely track the path of a typhoon brewing off Japan or of a zigzagging hurricane in the South Atlantic, follow the movement across the Sahara of severe sand storms or of violent thunderstorms over the European Continent, and gauge with unprecedented accuracy massive arctic ice formations.

These satellites are designed so that foreign nations can use either inexpensive ready-made or easy-to-build Automatic Picture Transmission (APT) sets to obtain daily weather forecasts directly from space. Sets are used today in sixty-five nations and NASA estimates conservatively that tens of thousands of human lives and billions of business-industry dollars have been saved just by storm warnings.

To learn more about nature's fickle characteristics, Argentina and Brazil have, since 1966, launched more than one hundred rocket-sounding missions synchronized with similar flights from various American sites, on a north-south line, as part of an Inter-American Experimental Meteorological Rocket Network. NASA is also cooperating with France in a weather satellite and balloon project called EOLE. The French National Space Commission built a small satellite that was launched by an American Scout rocket in August 1971 in conjunction with five hundred balloons carrying electronic packages emitting a distinctive signal. Global winds were studied as the satellite identified and located each balloon, deriving its path from a series of electronic interrogations and obtaining additional information on pressure, temperature, and balloon superpressure. To gather such information and to be able to carry out this type of project is important to the Global Atmospheric Research and World Weather Watch programs of the World Meteorological Organization and the International Council of Scientific Unions.

A dozen countries built ground terminals at their own expense during the 1960s to help NASA test and develop experimental communications satellites, such as Relay, Telstar and Syncom. From this cooperative beginning has evolved the International Telecommunications Satellite Consortium (Intelsat), now more than eighty nation-members strong.

From its constantly operating spacecraft, strategically positioned about 23,000 miles over the Atlantic, Pacific, and Indian Oceans, live TV coverage can be beamed, as we have seen, to hundreds of millions of people over most of the earth. Perhaps even more important is the payoff in direct economic gain to world commerce. Since each satellite can carry five thousand telephone conversations or twelve television programs, as well as tens of thousands of teletype circuits, they provide cheaper, more reliable long-range communications to all parts of the world.

In addition to Intelsat (see pages 56–59), a number of nations have domestic satellite systems planned or operating. For example, the United States has launched Telesats I and II for Canada (see Chapter 4) to provide telephone and color television service to all parts of that country from Vancouver Island to Halifax.

The United States and Canada are cooperating on development of a Cooperative Communications Technology Satellite to test, in a new frequency band, high-power satellite transmission to small American and Canadian terminals. Educational and medical information will be transmitted, after a 1975 launch, from this spacecraft in geostationary orbit. Canada is building the satellite and the United States is providing one experiment and the launch vehicle.

Exciting advances in education, made possible by the space program, promise a formidable weapon in the fight against ignorance. NASA soon will launch, for instance, a sun-powered satellite, ATS-F, with a thirty-foot dish antenna that will unfold in space. It will be used initially by United States government agencies in a number of experiments in communications development and environmental factors. Then it will be nudged further east in its synchronous orbit until it is in line-of-sight of an experimental ground station in Ahmedabad, India, where, it will remain in a stationary position.

From a few transmitting stations the Indian Government will beam educational television programs—focused initially on population control and agricultural improvement—to the satellite. ATS-F will retransmit the programs to many thousands of people in five thousand rural villages equipped with inexpensive Indian-made community TV receivers.

Some 600,000 direct-broadcast receivers will ultimately be set up centrally in Indian villages for audiences of up to several hundred people at each location. The eventual benefits of wide-ranging educational programs are enormous, and the system is remarkably economical. The cost of space television is about half of what an equivalent ground-based system would cost, for satellites do not need large earth receiving and transmitting stations and complex relay networks.

Brazil is also studying an educational system calling for

New educational tool: Artist's rendering of NASA's Applications Technology Satellite, which is planned to orbit over India and broadcast educational television programs to some 5,000 Indian villages.

direct satellite broadcasts. Brazil's problems are somewhat different from India's, for there is a distinct population imbalance caused by very sparse population distribution: 90 million Brazilians are spread over more than 3 million square miles, and there simply are not enough teachers to go around.

Through satellites, however, Brazil hopes to instruct students in rural areas on modern agricultural methods and to provide sufficient basic education to help overcome the school system's geographically caused deficiencies. Plans call for direct satellite broadcasts to about 150,000 schools, reaching 30 million people—about double the number of Brazilians now receiving schooling. Again, the potential savings are substantial: officials estimate that the cost of the spaceborne system would be one-fifth of the earthbound.

Other nations will watch the progress of the Indian and

Brazilian satellite educational efforts, for their success could lead, someday soon, to truly international programs. It is thus not inconceivable that there will be an absolute end to illiteracy or a time when all the world's peoples will speak a common language.

The annual cost of equipment for a truly global satellite educational system that broadcasts radio and TV programs to all countries has been estimated at about one dollar per pupil. Such a system could eventually be developed to permit thousands of programs to be broadcast simultaneously on different frequencies so that the viewer can select the subject he prefers. Educational TV broadcasts in India via communications satellite may show whether the method provides a solution to the developing nations' educational problems. A global communications satellite system might also exploit the vast potential of lifelong education.

Besides earth resources, weather, and communications satellite use, NASA cooperates with nations in wide-ranging scientific space programs. Foreign-designed, -financed and -built satellites are orbited by American rockets and the data harvest is shared.

NASA has launched and continues to launch instrument-packed vehicles for the United Kingdom, Canada, Germany, France, Italy, and the European Space Research Organization (ESRO). Scientists have gleaned much knowledge from them that is helping unravel old mysteries about the ionosphere, the magnetosphere, radiation belts, and other near-earth and space phenomena.

Of the many such projects planned for the next few years, one of the most ambitious is called Helios. Two German spacecraft, to be launched by United States boosters in 1974 and 1976, will record physical measurements within 28 million miles of the sun—closer than any spacecraft has been before. Of the ten experiments to be launched, seven will be provided by German scientists and three by NASA in cooperation with United States, Australian, and Italian scientists. Germany will pay the major share of the cost, estimated to exceed $100 million.

A joint NASA-ESRO scientific project of the 1970s, the International Sun-Earth Physics Satellite Program (ISEPS) is oriented toward the study of solar winds. ESRO will develop one satellite and the United States another, and both will be launched by a single rocket, perhaps in 1977.

Being readied for flight, this Highly Eccentric Orbit satellite was designed and built by the European Space Research Organization (ESRO) to study interplanetary physics and the high latitude magnetosphere. It was boosted into space by an American launch vehicle.

ESRO is also working on the preliminary design of a Highly Eccentric Lunar Occultation Satellite (HELOS), which will be devoted to X-ray astronomy and is to be launched in 1979.

On October 9, 1972, President Nixon announced that for satellite projects for peaceful purposes the United States would provide launch assistance, on a nondiscriminatory reimbursable basis, to other countries and international organizations.

A few months later the United States and the United Kingdom concluded an agreement providing for the latter's access to American launch capabilities on a reimbursable basis. The British Department of Trade and Industry will purchase appropriate rocket boosters and launching services from NASA for its own satellite projects.

The first British satellite to be launched is the X-4 technology research vehicle, which will be orbited in 1974 from the Western Test Range at Vandenberg Air Force Base, California.

American-British cooperation in space activities began in 1960 and has included British support of several NASA tracking stations, joint testing of experimental communications satellites, numerous sounding-rocket projects, lunar sample experiments by British scientists, and cooperative scientific satellite projects involving five launchings to date.

In addition to cooperating on satellite ventures, NASA solicits foreign proposals for experiments to be flown on NASA spacecraft. More than twenty experiment-packages—from France, the United Kingdom, Italy, The Netherlands, and Switzerland—have recently been selected on their merits through worldwide competition.

Foreign scientists thus have opportunities to participate in useful flight research, while NASA gains access to outstanding space science capabilities in the international community. Flight opportunities are made available at no cost to the foreign cooperators, who, in turn, make their experiments available at no cost to NASA.

Other countries without the resources for large-scale satellite efforts also can and do take active part in international space programs. Such participation is usually via sounding-rocket flights, which include smaller but scientifically significant payloads. Sounding rockets are principally used to study the earth's atmosphere above the limits of balloon flights (twenty miles) and up to the point of lowest satellite investigations (one hundred miles).

Sounding rockets are also used to determine the pressure, density, and temperature of the ionosphere, aurorae and airglow, solar flares, geomagnetic storms, trapped radiation fluctuation, and meteor streams.

To date, more than eighteen hundred sounding rockets have been launched by NASA at White Sands, New Mexico; Wallops Island, Virginia; Fairbanks and Point Barrow, Alaska; Fort Churchill and Resolute Bay, Canada; Thumba, India; Kiruna, Sweden; and Corou, French Guiana. And NASA continues to launch about seventy-five annually.

Other scientific packages have been launched by American rockets at facilities developed in Brazil, Argentina, and Pakistan, at locations suitable for research into special polar, auroral, and equatorial regions.

The Italian space experiment San Marco III, for example, in 1971 was launched into space by a United States rocket from a platform in the Indian Ocean off the Kenya coast. It measured local densities of the equatorial upper atmosphere in a zone high above earth that could not practically be reached from American launch sites.

The encouraging and growing international spirit of cooperation in space also extends to physical sites on earth. The continuing tasks of tracking, communicating with and acquiring data from the multitude of NASA's manned and automated spacecraft—from the tiniest satellite to the manned Apollo lunar landings—has required the extensive and intimate participation of twenty-one nations. Some twenty stations around the world are actively operated in support of programs like Skylab.

Perhaps nowhere is the true spirit of sharing the great wealth of knowledge obtainable from space more eloquently expressed than through the medium of the moon rocks and soil brought back to earth by the twelve astronauts who physically explored the lunar surface.

Some of the world's foremost scientists in eighteen nations have analyzed and continue to analyze these priceless geological samples. These investigators, selected on the merits of their

From an ocean platform near Malindi, Kenya, an Italian-built San Marco satellite, designed to make detailed observations of the earth's upper atmosphere, was launched by a United States–built, four-stage Scout rocket.

proposals for the studies they were best qualified to perform, and given no American financial support, are carrying out a full range of physical, chemical, mineralogical, and biological experiments on the lunar samples. They are joined in their researches by more than 150 colleagues in universities and research centers across the United States. Nations everywhere are thus afforded the unique opportunity to share in one of man's truly great endeavors, and the results of their detailed studies—some taking several years to accomplish—are in turn shared by the international scientific community.

Foreign scientists and engineers also are encouraged to study and work on space-related projects at American universities and NASA Centers, and by late 1973 more than one thousand individuals from forty nations have done so. In many cases the participants in these programs have returned to their respective countries to serve as the nucleus for the development of national space organizations and programs.

NASA leaders have also toured many of the space capitals in the Western world—London, Paris, Bonn, Madrid, Brussels, The Hague, Ottawa, Canberra, and Tokyo—to brief officials and scientists on and to elicit active involvement in future space missions.

NASA administrator Dr. James C. Fletcher has said President Nixon is especially interested in carrying foreign experiments and experimenters on the space shuttle. "Anything involving manned flight has worldwide interest," he says. "This will be particularly true of the shuttle. Everyone wants to use it, and we envision widespread participation.

"The President is very much interested in international aspects of the shuttle and believes space should be used for all mankind. He would like to see a foreign astronaut on one of our flights, and the shuttle could well be the vehicle for this."

"The shuttle represents an ideal opportunity to cooperate with other advanced nations of the world," says Senator Frank E. Moss, chairman of the Senate Aeronautical and Space Sciences Committee. "The Europeans, we know, are already busy planning for active participation in the program. We expect the Japanese to become involved. I wouldn't be surprised if Eastern European countries express interest. And, it is entirely conceivable that we will cooperate with the Soviets on the shuttle. I would hope such cooperation will be encouraged.

"I believe it would be disastrous if we couldn't move forward with international cooperation," says Senator Moss. "One thing we must do is keep space open for the worldwide exchange of knowledge."

There is already considerable overseas interest in the shuttle program. With European funding in the early 1970s British, French, and West German industrial firms worked with NASA's prime contractors in design studies for the shuttle. The European Space Conference also independently funded studies for designs of a vehicle that could fly to and from different orbits to service the shuttle.

On September 12, 1973 the European Space Research Organization (ESRO) signed an agreement with NASA to develop on its own a "sortie laboratory"—to fly with NASA's shuttle—which will have two elements: a pressurized manned laboratory module, and an external unpressurized instrument platform, or pallet, suitable for conducting research and applications activities on shuttle missions between orbits—lasting from seven to thirty days.

Both the laboratory module and pallet will be ferried to orbit in the payload bay of the shuttle orbiter, and will remain attached to the shuttle throughout the mission. As it is now envisioned, a staff of up to six scientists and engineers will work in the sortie lab, but eat and sleep in the shuttle while in orbit. They will be able to pursue experimental activities in the laboratory module working in a normal shirtsleeve environment.

Germany, Italy, Belgium, Denmark, France, the Netherlands, Switzerland, the United Kingdom and Spain are the nations which have agreed to participate in development of the sortie lab, although other countries may join later. It has been estimated that the project will cost some $400 million, to be funded by participating European nations. Tentative plans call for a flight unit of the vehicle to be delivered to NASA in 1979.

"Because the shuttle will substantially reduce the costs of doing business in space, cutting the costs of both launch vehicles and payloads, it demands the attention of other nations beside ourselves," explains Arnold Frutkin, NASA assistant administrator for International Affairs. "The shuttle offers volume, weight, preparation, environmental and cost advantages which no conventional launcher system could match. It must therefore become the preferred means of getting into space for any nation wishing to use space.

"The significant aspects of all our efforts," says Frutkin, "are the cost savings, the scientific and technological benefits, the impetus to space research, and the strengthening of working relationships with the world's scientific communities."

Through its international programs during the 1960s and early 1970s, NASA established a broad base of institutions, facilities, competence, and patterns of cooperation from which it can move confidently in the future, in concert with all nations.

NASA is at present engaged in a major effort to increase international cooperation throughout the 1970s and the rest of the twentieth century, by extending its activities with Western nations to include participation in the development and use of such major new space systems as the shuttle, and in the experimental development of space technology applications.

The objective, as Frutkin pointed out, is to bring about a greater sharing of the costs and the benefits of the exploration and utilization of space.

"Space may not offer solutions to all the problems of the world," says Dr. Fletcher, "but it does offer us a new source of hope. And we get more international cooperation on space ventures than we do on anything else."

"Views of the earth from space have shown us how small and fragile our home planet truly is," President Nixon observed. "Perhaps ultimately men of all nations will work together in space . . . and through their activities help humanity unite in peace on its planet earth."

O
Chapter 16
orbital comradeship

We met at the Elbe [during World War II] and we can meet over the Milky Way.

—YEVGENY YEVTUSHENKO
SOVIET POET

On May 25, 1972—eleven years to the day since President Kennedy's designation as a national goal the landing of American astronauts on the lunar surface—newspapers everywhere banner-headlined the announcement, datelined Moscow, of equal magnitude concerning space flight: U.S., RUSSIANS AGREE TO JOINT SPACE MISSION.

The story described how President Nixon, on an official state visit to the Soviet Union had signed an agreement with Soviet Premier Aleksei Kosygin in which astronauts and cosmonauts would go into earth orbit together.

The President and the Premier drank champagne toasts, and

the historic proclamation was hailed everywhere as a dramatic advance. Not only did it team up the two space powers, but it was also a giant positive stride toward a permanent easing of international tensions, and toward the truly cooperative peaceful use of space for universal benefit.

Behind the headlines but not covered in any of the news dispatches, lay more than a decade of hard, patient, often frustrating, endlessly detailed groundwork, without which the signing in Moscow would not have been possible.

It could be said that the joint Soviet-American move toward space actually began in a highly dramatic manner—with the launching on October 4, 1957, of Sputnik I. That single launching set the two nations off on a fiercely competitive race of science and technology that for all practical purposes halted abruptly when astronaut Neil Armstrong stepped onto the lunar surface in July 1969. In effect, so long as the moon race was on, there were no real breakthroughs in large-scale cooperative ventures between the two superpowers.

There were, however, some encouraging signs over the years, one of the most promising being the adoption on December 20, 1961, of United Nations General Assembly Resolution No. 1721, which cleared the way for initiation of a comprehensive UN-sponsored program of space cooperation. Within this framework, the late Dr. Hugh L. Dryden, then deputy NASA administrator, and Soviet Academician Anatoly A. Blagonravov negotiated a series of bilateral space accords about exchanging weather data, mapping the earth's magnetic field and exchanging magnetic observatory data, participating in satellite telecommunications experiments with the American satellite Echo 2, and exchanging information on space biology and medicine.

Such agreements were less important in their physical results, which were minimal, than in the effect they eventually had on building confidence for much broader future space cooperation. At least some seeds were planted.

There were, of course, many hard, practical reasons for pooling space capabilities and expertise, some of which were cited in a staff report on Soviet Space Programs prepared for the Senate Committee on Aeronautical and Space Sciences by Joseph

G. Whelan, a specialist in Soviet affairs at the Congressional Research Service of the Library of Congress:

> The universality of space, its sheer magnitude and potentiality for benefiting humanity, made it a vast sharable resource. The enormity of the tasks in exploring space, moreover, seemed to impose upon man the necessity for cooperation in some form, especially in far-reaching, expensive experiments. The universality of science also imposed upon the fraternity of world scientists a shared professional concern for the universe and an innate intellectual curiosity about what made it work.
>
> In a practical way, all nations had a shared interest in space, and in specific instances, as in allocating radio frequencies for space communications and forecasting weather conditions on a global scale, individual interests could be achieved only through cooperation with other nations for the same common good. . . . Space exploration was an international activity requiring such facilities as tracking stations and agreements on the rescue and return of astronauts and space equipment. Thus, by the nature of the enterprise, some form of cooperation was an objective necessity.
>
> The rationale can be reduced to this simple line of reasoning: Space exploration is expensive. Why duplicate research that leads to waste in money and resources and does not necessarily guarantee positive results? Why not conduct joint space experiments and even establish a division of labor for planetary and interplanetary research? Why not set up a procedure for the exchange of information? The result would be savings in costs and resources for both space powers and progress in space exploration. The presumption is that the Russians and the Americans have a shared economic interest in cooperation of this nature.
>
> The case for space cooperation was, therefore, formidable.

Summarizing, Whelan wrote:

> Whether space exploration can ever be a truly integrated international concern or . . . will become polarized along political lines depends in large measure upon the willingness of the major space powers to share their bounty and their unique skills with the rest of humanity.

But, most of all, it depends upon their willingness to agree
. . . on the terms of reference for meaningful and productive
bilateral cooperation in space; for it is upon this foundation of
agreement that must be built the larger structure and superstruc-
ture of international space cooperation.

On the significance of such cooperation *The New York
Times* editorialized: "Born in the mad competition for status
characteristic of the cold war, manned and unmanned space
research has taught both sides how puny are man's resources in
facing the mystery and challenge of the universe. As that lesson
has sunk in, both sides have come to understand the advantages
of cooperation as against useless and wasteful rivalry."

The United States's enthusiasm for openly sharing space has
long been established, and was even written into the Space Act of
1958 as a matter of national policy. In addition to cooperating in
space programs with nations all over the world, NASA early took
the initiative for suggesting some meaningful form of space
cooperation with the Soviet Union. And the United States is
committed in the decade of the 1970s to expand such cooperation
with Russia and all other countries.

Dr. George Low, deputy NASA administrator, summed up
the American attitude when he said, "We are striving to cooperate with the Soviets in the exploration of space because we both
live in a vast universe that must be explored, where important
new knowledge is to be gained, for the benefit of all men,
everywhere. It is in the United States's interest to cooperate with
all nations, including the Soviet Union, to share our resources in
gaining this knowledge, and to share the results for the betterment of mankind.

"And even though . . . 'competition' and 'cooperation,' at
first glance, appear to be incompatible, the USSR and US space
programs are indeed developing along both of these lines: we
were competing in technology, when they first put a man in
space, or when we first landed a man on the moon. We are
cooperating, for the sake of science, when we exchange data from
space."

Negotiations between the two nations had been under way
since 1961, but the space venture between America and Russia
did not actually begin until August 18, 1966, when the first

meteorological satellite data were received from the Soviet Union. They consisted of cloud analyses hand-drawn from information in television and infrared photographs. Three weeks later the United States began transmitting to Moscow television cloud photographs and cloud analyses.

There was also some progress in magnetic-field mapping and communications, in the exchange of information on space biology and space medicine, and in cooperation in tracking and in data acquisition. The effort was not, perhaps, wholehearted and the results were far from what might have been expected, but it is not entirely fair to assess Soviet participation in joint space ventures during the 1960s, in the words of a NASA official, as "disappointing and discouraging." Rather, it would be more appropriate to appraise the progress made, in Arnold Frutkin's words, as "very slow, but heartening."

Following America's first manned lunar landing, the spirit of Soviet-American cooperation improved measurably, finally paying dividends for the untiring, painstaking early efforts of such dedicated men as the late Dr. Dryden. In January 1971, for example, a team of high-ranking American space officials representing the President met in Moscow with their Soviet counterparts and prepared a list of specific areas for exchange of data between countries.

A joint working group was set up to organize space meteorology experiments designed to advance the knowledge of temperature sounding from satellites and microwave measurement of precipitation zones, ice conditions, and sea-surface roughness and temperature. Another group provided for coordination of meridional sounding-rocket networks in both hemispheres and began an exchange of operational and scientific meteorological data.

American and Soviet specialists are teaming in studying the natural environment, to define coordinated experiments in remote environmental sensing of vegetation, geology, and the oceans.

A joint group on space biology and medicine has exchanged highly detailed data and results from the manned Soyuz/Salyut and Apollo programs, and has taken the steps necessary to augment the information flow between the nations.

There has also been exchange of lunar samples, beginning

United States–Soviet cooperation through mutual space interests was demonstrated when Soviet Academician A. P. Vinogradov, center, presented NASA's Lee Scherer, second from left, with lunar samples gathered by the unmanned spacecraft Luna 16. Earlier, the United States had given Soviet scientists lunar samples from material brought back to earth by the Apollo 11 and 12 astronauts.

with the swapping of three grams of material obtained by the unmanned Luna 16 for three grams each of the lunar soil brought back by Apollo 11 and 12 astronauts. These relatively small amounts were sufficient for detailed comparative scientific examination. Exchanges of lunar rock samples continued throughout the Apollo and Luna programs. Soviet and American experts have been working on a common system of lunar coordinates and a program for compiling a complete lunar map.

Joint American-Soviet ventures have been extended to go far beyond earth-moon space: the cooperation now reaches, literally, to Mars and Venus. Under terms of new agreements drawn up by leading interplanetary scientists of both countries, NASA is providing the Soviet Academy of Sciences with maps and photos taken by Mariner spacecraft of two Mars landing regions of interest to the Soviets; Mars ephemerides from ground-based-radar data; radar measurements of Mars; and results obtained on Venus during the Mariner Mercury-Venus flyby in 1974.

In return the Soviet Academy provides NASA with information on Martian landing sites in the forthcoming Viking program collected by Soviet spacecraft; matériel on atmospheric parameters and the surface of Mars; data obtained by Venera 8 on the atmosphere and surface of Venus, and radar measurements of Venus.

With the aim of strengthening the legal order in space and developing international space law, the United States and the Soviet Union are studying the legal aspects of space, for both are keenly interested in encouraging international resolution of problems involving international law in the exploration and peaceful use of outer space. This relates to the continuing effort, in the context of the United Nations Outer Space Committee, that has resulted in a treaty on principles governing national and state explorations and uses of outer space, and in an agreement on assistance to astronauts and the return of matériel.

The possibility of a cooperative American-Soviet manned mission was initially proposed in the fall of 1963, when President Kennedy suggested it as part of an integrated moon program. The Russians at the time, however, were not interested.

Not until more than seven years later, in October 1970, did the Soviets agree to a NASA proposal to consider the possibility of designing spacecraft to permit rendezvous and docking in outer space. The discussions were extended to decide on a test mission in earth orbit between spacecraft of both nations in the mid-1970s.

A series of meetings and planning sessions that continued into the next several months ended with the Nixon-Kosygin signing that committed the nations to a joint manned space venture. An American Apollo will link up in space with a Soviet Soyuz. The Apollo will be a modified version of the familiar command and service modules used in all manned lunar landings—in fact, the craft was originally built for a moon shot.

The smaller Soyuz (pronounced sah-YOOZ), which since 1967 has been the prime Soviet manned spacecraft, has three essential units: an orbital module, at the forward end, used by the crew for work and rest in orbit; a descent module, with main

Brotherhood in space: An artist depicts how a United States Apollo and a Soviet Soyuz spacecraft will be mated in earth orbit in 1975 for what will be history's first international space mission. (Courtesy Space Division, Rockwell International)

controls and crew couches, used during launch, descent, and landing; and an instrument module, to the ship's rear, containing power, communications, propulsion, and other subsystems. Because of its smaller size, Soyuz is essentially a two-man ship, hence only two cosmonauts will fly on this joint mission, but there will be three American astronauts in the considerably larger Apollo.

First, Soyuz will be launched from the Soviet space base at Tyuratam, in Central Asia; once it has reached orbit the cosmonauts will level their craft about 167 miles above earth.

Seven and a half hours later, around the world at the Kennedy Space Center, Apollo will blast off.

An hour after entering orbit about 124 miles up, the Apollo

command and service modules will separate from the second stage of the Saturn 1B launch vehicle, turn around, and link up with a custom-designed docking module.

After routine checkout the astronauts will begin a series of maneuvers that during the next several hours will place them in the orbit with the Soviets. The two craft will soon dock in space, using techniques developed and perfected in the last decade by both nations. On Apollo's fourteenth revolution of earth, nearly a day after launch, it will approach Soyuz, and the two will bind together with common docking mechanisms.

The present flight plan calls for the Americans to visit the Soyuz craft for several hours, entering through a docking module. They will carry voice-communications equipment and a television camera. NASA's Dr. Fletcher says this part of the mission will be the most visible Soviet-American cooperative effort ever, "since it involves cosmonauts and astronauts working together on a very complex mission while the whole world is observing on television via satellite relay."

After the initial visit an astronaut will accompany a cosmonaut to the Apollo. The two ships will remain docked for about two days, and Apollo will remain in orbit for up to eleven days more on an earth-resources survey mission. Definite objectives of the mission include:

- testing of rendezvous and docking techniques involving two spacecraft, leading to the day when a common docking assembly can be used to join the orbiting space vehicles of any two nations;
- verification of the techniques for transferring the astronauts and cosmonauts from one craft to the other.
- the carrying out of a program of experiments including ultraviolet absorption, biological interaction, microbial exchange, multipurpose furnace, and artificial solar eclipse. Selections were made from 145 proposals received by NASA in response to invitations issued to American and foreign scientists. All results will be made available to the international scientific community.
- gaining of experience in the conduct of joint flights, including the rendering of aid in emergency situations via rescue missions; the exchange of cosmonauts and astronauts will be a realistic rehearsal for a procedure that could be used to rescue crew members from a distressed space ship and transfer them to another craft.

It still is difficult to conceive, but within a decade—by which time advanced transportation vehicles such as the space shuttle and space tug, as well as future-generation Soviet ships, will be routinely zipping to and from space—dozens of astronauts and cosmonauts undoubtedly will constantly be either in flight or on space stations. Such intensified activity increases the laws of probability coming into play. Manned spacecraft will eventually be in distress, and the only vehicle in a position to help it will be one of another nation. The ancient tradition of the sea—ship speeds to the help of a distressed ship of any registry—would thus be extended into space.

Although the Apollo-Soyuz flight will not be made before mid-1975 at the earliest—and no major technological breakthroughs are necessary—serious preparations were being made by both nations well before the May 1972 agreement.

A number of basic technical problems must be solved, primarily the compatibility of the two different sets of spacecraft hardware. The space pilots and flight controller also need considerable time to learn the techniques, systems, and idiosyncrasies of each others' machinery and methods. And, of course, there is the language factor—astronauts and engineers in the United States are taking cram courses in Russian, and the Soviets are studying English.

The American crew will consist of Apollo veteran Tom Stafford, commander; rookie Vance Brand, command module pilot; and Donald "Deke" Slayton, docking module pilot. Slayton was one of the original seven astronauts selected in 1959 for Project Mercury (the nation's first manned space program), but he has never made a space flight. A cardiac irregularity grounded him for a number of years before clearing up and enabling him to return to flight status.

The only significant new piece of spaceware—the Apollo and the Soyuz, long since built, needed only modifications—is the docking module. Under a NASA contract it is being designed and developed by Rockwell International's Space Division of Downey, California, which was also the builder of Apollo.

The docking module—4.7 feet in diameter, 9.7 feet long, and approximately 3500 pounds—is a cylindrical structural and mechanical adapter to enable the Apollo command module and the

Soyuz spacecraft to dock, and to facilitate crew transfer between the two spaceships. One end of the docking unit will be equipped with an Apollo-type docking system, while the other end is to have a newly developed system that will permit docking with Soyuz. The docking mechanisms will be of universal design, adaptable to the future spacecraft of any nation.

It will also serve as an airlock for "atmospheric adaptation" during crew transfer—a kind of decompression chamber, necessary because the Soyuz atmosphere is 80 percent nitrogen and 20 percent oxygen, and its sea-level pressure is 14.7 pounds per square inch (psi); with 100 percent oxygen inside the cabin, Apollo has a low-pressure, 5-psi atmosphere.

NASA officials have indicated that the United States and the Soviet Union will spend approximately equal amounts on the mission, with the former's share estimated at about $250 million.

"What is of tremendous historical significance is not the physical mission itself, which does not represent any state-of-the-art advances in technology," says Dr. Fletcher. "It is, rather, the cooperative effort and all the exercise of good will that must precede this space maneuver. And what is more important still is the breakthrough this one scheduled test could provide in the direction of future cooperation. Apollo-Soyuz is not the culmination, but a very major milestone in a long period of negotiations with the Soviet Union on joint space ventures.

"I think the first mission of the U.S. and Russian astronauts hardly will be the last. It is our hope that this first mission is the precursor of future joint manned and unmanned efforts. Such cooperative programs will enable both countries to better serve all mankind with continued vigorous efforts to expand our understanding of science and development of new technology for better life on earth."

The Soviet-American manned space docking agreement may eventually prove to be symbolic of a mutual awareness of the planet's problems that have immense importance for people everywhere, whatever their political orientation. Ultimately it may symbolize awareness that earth itself is a spaceship in which the fates of all earthlings are linked.

○
Chapter 17

the new high ground

In 1972 the Soviet Union launched into orbit a satellite, designated Cosmos 462, from the Tyuratam rocket base but offered no public announcement as to the ship's payload or its purpose. Within hours it moved close to Cosmos 459, launched four days earlier, and exploded into thirteen pieces. Both satellites were destroyed 150 miles above earth.

Diagnosis of this type of mission was made in a detailed special Congressional study, *Soviet Space Programs,* made by the United States Senate Committee on Aeronautical and Space Sciences, which said, "It now seems a reasonable inference that the Soviet Union has actively pursued and possibly perfected a system which is capable of reaching a co-orbit with another satellite which is uncooperative, making some kind of an inspection, and if deciding it is hostile, destroying it."

In other recent tests Cosmos interceptors exploded near target spacecraft flying at higher altitudes—on spatial courses similar to those flown by American military communications and navigation satellites.

It is known too that the Soviets are testing hardware components that could be needed if a nuclear deterrent force remains in orbit for sustained periods. The Senate report noted that since 1966, Russia has launched at least sixteen spacecraft to test this "Fractional Orbiting Bombardment System" (FOBS). This system, according to the report, can place a nuclear warhead in orbit and bring it down on a ground target after less than one full revolution of the earth.

Such a system works this way: from well-fortified launch sites deep in the Russian interior, giant rockets boost spacecraft armed with nuclear warheads into orbit above earth. From such a vantage point these warheads could be dispatched—at hypersonic speeds through space and back into the earth's atmosphere and toward virtually any target on the planet—before any retaliatory action could be taken. An equally grim possibility would be the stationing around the world of a series of lethal satellites for use in global nuclear blackmail.

The probability of the Russians' ever using such awesome weapons against foreign nations—like that of their using nuclear bombs or missiles—is remote. But this does not alter the fact that *they are developing these capabilities!* It is the old arms race taken to the ultimate environment, space.

While the United States and the Soviet Union are striving for international cooperation in space and are working on a joint Apollo-Soyuz flight for 1975, it has not changed the hard fact that Russia is spending far more on her military space programs than America. In 1971, for example, there were only thirty-one launches in the entire United States space program, and the majority of these were made by NASA for peaceful purposes. In contrast, the Soviets orbited eighty-three payloads that year, of which fifty-nine were for military missions.

At a 1972 Congressional hearing, Dr. Robert C. Seamans,

former secretary of the United States Air Force, said, "The best information available indicates that they [the Soviets] plan to spend more on space this year than both the Department of Defense and NASA combined. I believe that the priority given to our space efforts should reflect the fact that our national security could be seriously jeopardized if another nation should move very far ahead of us in space technology."

Statistics support Dr. Seaman's comments. Recent intelligence reports give the Soviet Union about 610,000 full-time research-and-development scientists and engineers, while the United States has only 570,000 (the first time the Soviets have had an edge). Soviet funding for research-and-development programs has leaped from $3.5 billion fifteen years ago to more than $21 billion. Since the Apollo landed on the moon, the United States has drastically reduced its space budget.

United States Congressman Olin E. Teague, chairman of the House Subcommittee on Manned Space Flight, warns, "I am convinced that only so long as the United States is a major space power will the free use of space be available to all mankind. Space today is the 'high ground' militarily. To abrogate the field to a potential enemy is to court disaster."

Responsibility for the military aspects of space rests primarily, of course, in the Department of Defense: The Air Force has been launching satellites for national security missions since 1958—in fact, America's first satellite, Explorer I, was launched into orbit the night of January 31, 1958, by a Redstone booster originally developed for the United States Army. This same rocket started astronauts Alan Shepard and "Gus" Grissom on their historic 1961 Mercury suborbital flights. Modified Air Force Atlas rockets sent John Glenn, Scott Carpenter, Wally Schirra, and Gordon Cooper into earth orbit, and Air Force Titan missiles were used to boost astronauts into space throughout the Gemini program which led to Apollo.

The Air Force's Agena vehicle was the first spacecraft to achieve circular orbit, the first to propel itself from orbit to orbit, the first to send a spacecraft on successful Mars and Venus flybys, and the first to rendezvous and dock with another spacecraft.

The Defense Department's contributions to the national space program are continuing through the 1970s. NASA has selected a modified version of the Air Force's Titan III D launch vehicle and a Centaur upper stage for use in a number of space tasks, such as the Mars-Viking mission.

For its own purposes, however, the Defense Department's strategic philosophy, simple but sound, has been to view space as the natural progression to a new environment, a new medium, one in which to continue the job of protecting national interests. Space is merely a step beyond land, sea, and air parameters, as fundamental as high ground to an infantryman.

"NASA has a space mission, the Defense Department does not," says Grant Hansen, former assistant secretary of the Air Force for research and development. "The Defense Department mission is national security through such submissions as strategic forces, tactical forces, mobility forces, intelligence and security, training, and others. In the Air Force the fact that something can be done in space is a necessary but not sufficient condition. Space must be the *best* way to accomplish an essential function for a military mission."

In the past fifteen years the Defense Department has launched hundreds of unmanned satellites into space for various purposes. "Since the beginning of the space era, the Air Force has taken the lead in developing space systems to improve our national security," says Dr. Seamans. "The Air Force has responsibility for the development and launch of military space boosters for the Department of Defense and is the principal defense agency for space development programs and projects. This role as Defense's primary space agency includes the tasks of supporting the Army, Navy, and Atomic Energy Commission space requirements. It also includes working with NASA on Project Apollo and launching NASA satellites as necessary. The Air Force has launched about two-thirds of the more than seven hundred payloads the United States has launched into earth, sun, or lunar orbit since 1958.

"We have space-based systems for early warning, communications, charting, and geodesy, some in collaboration with NASA. These systems do not pose a threat to the security of other nations. They are used to provide information that enhances our deterrent capability, not to carry weapons that might be used offensively."

Hansen describes some of the distinct advantages of using space for military purposes: "Because of the tremendous area of earth coverage from a satellite, its capabilities are highly superior for data-gathering functions requiring broad coverage, such as cloud cover and meteorology, earth-resources surveillance over land and water, communications relay, and detection of missile launches.

"Space is uniquely capable of wide-area coverage for detecting and characterizing missile launches. Satellites viewing a tremendous volume of near-earth space without atmosphere provide a unique capability for nuclear burst detection, to monitor treaty compliance, and for solar-radiation data-gathering to predict radio-communications performance."

Space is actually used for defense missions in a variety of ways, says Hansen: "Satellites may be used to sample and readout data from airborne or surface sensors, again taking advantage of the unique simultaneous 'line-of-sight' coverage from space to a very large area of the earth. A reverse of this process is the navigation satellite system in which sensors on the ground, on the sea, or in the atmosphere can use satellites as objects to be sensed in determining the accurate position of the sensor. The wide area of earth coverage and the stability of satellite orbits are the unique space capabilities which make navigation satellites a best way to perform a necessary function."

In the field of early-warning satellites the Air Force is developing [Defense Department] space systems for the purpose of detecting ICBM [intercontinental ballistic missiles] and SLBM [sea-launched ballistic missiles] launches and to report atmospheric nuclear explosions.

These early-warning or photographic reconnaissance satellites serve two complementary missions: one takes panoramic photographs from earth orbit to detect evidence of new construction and such installations of military interest as airfields, missile sites, and strategic targets; the second is a "close-look" satellite used to obtain high-resolution photos of specific installations. It is believed that the latest camera and sensing systems aboard the spacecraft are capable of photographing foot-high objects from orbital altitudes of one hundred miles or more.

Both types orbit the earth once in about ninety minutes, passing over the total global surface of the planet twice daily, once in daylight, once in darkness.

The United States has also developed an early-warning satellite system for placement in geosynchronous orbit about 25,000 miles above earth. One such Defense Department craft is over the Indian Ocean to monitor land-based Soviet missile sites; another, located over Central America, monitors for SLBMs in the near-Atlantic and Pacific. Together they provide increased assurance of detecting an intended surprise attack by either ICBM or SLBM.

They are also included in a Department of Defense network that includes information obtained from such other sensors as over-the-horizon radar and the Ballistic Missile Early Warning Systems (BMEWS). The importance of maintaining an integrated ground-airborne-spaceborne network is that, despite the Soviet Union's vast and growing arsenal of nuclear missiles, emplaced only thirty minutes from American cities, the Defense Department is fairly confident that a surprise attack like that at Pearl Harbor in December 1941 is almost impossible.

Actually, orbiting reconnaissance and nuclear-detection satellites tend to keep the world's major powers "honest." By providing an up-to-the-minute strategic weapons inventory, they allay fears by eliminating unknowns and actually can contribute meaningfully to international arms-control agreements, which would probably be impossible without them.

"In an effort closely related to the early-warning mission," says Dr. Seamans, "we are continuing our development of satellites for the Defense Communications System. The initial space system has been operational for several years."

Military communications satellites fulfill strategic as well as tactical missions. Strategic missions tie satellites to a group of fixed earth stations, hence communications can be established over any path on which two or more stations are mutually "visible" to orbiting spacecraft. The tactical system permits communications between satellites and a variety of earth stations that may include mobile ones such as those in aircraft, ships, automobiles, or even properly equipped infantrymen.

How effective are communications from space for military purposes? During the Vietnam war, satellites regularly transmitted high-speed digital data from South Vietnam to Washington. Within minutes after processing, high-quality reconnaissance photographs of battle zones were available to Pentagon analysts, via the Defense Department's Initial Defense Satellite Communications System. The revolutionary impact on strategic planning is obvious.

The Air Force's experimental Tactical Communications Satellite (TACSAT 1) has shown that spacecraft of this type can open a new era of communications for military and civilian communities—as it did in 1969, when it provided, for Alaskan TV stations, live coverage of the first moon landing. Alaska thus had the first live coverage of any national event.

The peacekeeping potential of orbiting spacecraft is exemplified by NATO's communications satellite system, which consists of two satellites in geosynchronous orbit about 23,000 miles over the Atlantic. Launched by the United States from Cape Kennedy, they are designed to insure rapid and secure communications among NATO members.

The Department of Defense is working on systems development. One is the Air Force Satellite Communications System (AFSATCOM), which will "provide a communications capability via satellite to satisfy high-priority Air Force requirements for command and control of forces." Operational testing was scheduled to begin in 1974.

The Fleet Satellite Communications System (FLEETSAT-COM), whose first launch is planned for late 1975, will provide an operational, nearly global satellite communications system to "support high-priority communications requirements of both the U.S. Navy and the Air Force." The space segment of the program will consist of placing four satellites in geosynchronous equatorial orbit. The spacecraft will have about thirty voice channels, twelve teletype channels, and a design life of five years. FLEET-SATCOM's ground segment will consist of communications links among designated and mobile users, including major United

States Navy craft and select Air Force and Navy aircraft, as well as global ground stations.

The Navy pioneered in the use of space for navigational purposes and has developed a Transit Improvement Program (TIP), which includes a study to determine the best way to reduce the interval between fixes, using Transit satellites to provide more frequent update for earthbound forces.

There are also plans for a 1974 launching of an experimental navigation satellite to study space "propagation phenomena" on various techniques for navigation signal modulation, as part of the major joint-service navigation space experiment for testing and demonstrating satellite navigation technology and capability planned for the 1977–1979 period.

The Department of Defense uses space for the collection and quick dissemination of weather information, and has also developed a new meteorological data system, the Defense Meteorological Satellite Program (DMSP).

"The meteorological aspects of DMSP were designed under a total-system concept in which not only sensors, but communications and ground processing facilities were developed with the primary objective of providing maximum responsiveness to the decision-maker, whether supported by a tactical field weather unit, or from a centralized weather facility," explains John L. McLucas, Secretary of the Air Force.

"Weather data are very perishable with time. The DMSP system has been designed to provide decision-makers with weather data within minutes of its collection in space. The space segments of the DMSP system consist of infrared and visual sensors; the infrared sensors' products are images of the earth and its atmosphere that are representative of their temperatures rather than their brightness, while the visual sensors detect the brightness of reflected solar illumination."

McLucas says that monitoring storms, including typhoons and hurricanes, on a global basis is one of DMSP's major applications, and that data gathered by the system's satellites are being made available to the public to gain maximum use of "this national resource."

All of these spacecraft systems—reconnaissance, communications, navigation, meterorological, and others or parts of them

—will eventually be carried to orbit by the space shuttle. This will greatly decrease the cost of today's individual launches, and launches whose booster rockets are discarded in the ocean. Launch costs alone for a Titan IIIC vehicle, for example, are in excess of $20 million.

"An intelligent visitor from another planet would probably think we were pretty dumb as he sees us throwing away these expensive machines after a single use," says Grant Hansen.

"For several years the Air Force and NASA have been studying ways to reduce the cost of space operations," adds Dr. Seamans. "We have concluded that the most promising way to realize the significant costs savings is through the use of reusable launch vehicles. Savings could also be realized from being able to recover satellites or conduct in-orbit maintenance on systems that malfunction.

"President Nixon's [January 1972] decision to proceed with space shuttle development initiated a program which holds great promise for scientific and technological advances in the interests of the nation and all mankind.

"We are also interested in its potential as a means for performing our military mission more effectively and economically. The Air Force role in the program is to provide NASA data to help assure that the shuttle will be of maximum utility to the Department of Defense, and we are pleased that the . . . vehicle is [being] configured to meet potential DoD needs. We will continue our close coordination with NASA as their development program proceeds."

Gen. Samuel C. Phillips, Commander of the Air Force Systems Command and former Apollo program director for NASA, views the shuttle . . . "as a dramatic step toward a fundamental change in the environment of space operations—a quantum jump between past space spectaculars and a routine and more effective employment of space in the future."

Present projections are that perhaps 30 percent of the shuttle's annual payload, once it has become operational in the 1980s, could be devoted to national defense missions. After extensive analysis the Air Force developed a tentative mission model designed to meet the Defense Department's needs in the 1980s and 1990s. The annual average number of Defense Department

Versatility of the space shuttle, which will deploy Department of Defense satellites as well as those of NASA and other agencies, will cut the costs of operating in space sharply. (Courtesy Space Division, Rockwell International)

shuttle payloads reflected in this model is about twenty, representing approximately today's level of effort.

Working closely with NASA to define a shuttle system that will meet Defense Department and NASA needs, the Air Force Space and Missile Systems Organization (SAMSO), headquartered in El Segundo, California, is providing the military requirements. These include minimum launch azimuth constraints to permit maximum mission flexibility for a number of Defense Department space missions, the ability to inject payloads into a variety of orbits, to change orbital parameters, and return from orbit under relatively unconstrained conditions.

After a mission the shuttle should be capable of hypersonic crossrange maneuvering to a predetermined landing site. And since Defense Department missions could require that payloads be transferred from low- to high-energy orbits—for example, from a circular orbit a few hundred miles above earth to a geosynchronous orbit 22,000 or more miles high—the propulsion stage for making this transfer must be considered part of the shuttle payload.

General Phillips says the Air Force is supporting the shuttle program where possible with unique Defense Department experience and expertise obtained during more than fifteen years of space flight. The Air Force also is extensively qualified to operate, test, and maintain large aircraft and is experienced in maintenance, refurbishment, and ground support of several missile systems and many aircraft—all of which is applicable to the shuttle.

"In regard to space tugs," says Hansen, "over half of the Defense Department's payloads projects for the future go to high-energy orbits which require an orbital-transfer stage in addition to the space shuttle orbiter. NASA has similar needs for high-energy missions. Just as it is a design goal that a single shuttle configuration will satisfy both military and other needs, it is intended that standard space tugs will meet the combined needs of NASA and the Defense Department, as well as other government, civil, and international users."

The Air Force also has a Space Test Program that provides launch vehicles, spacecraft, payload integration, and launch services to support various space-technology efforts of special interest to the Defense Department and peculiar to military needs. These include such projects as advanced space guidance, advanced space power sources, satellite secondary propulsion, and technology for improved satellite performance, reliability, and survivability.

"In looking to the future," Dr. Seamans sums up, "we must continuously assess our military requirements against available technology and fiscal considerations. Consistent with our nation's peaceful objectives and our commitments with respect to activities in space, we must make appropriate use of space systems where they can best help us accomplish our military tasks."

President Nixon has said that "defense technology has helped us preserve our freedom and protect the peace. Space technology has enabled us to share unparalleled adventures and to lift our sights beyond earth's bounds. . . . America must continue with strong and sensible programs of research and development for defense and for space."

Part Five

the
spinoff

○
Chapter 18

benefits at home

In the early 1970s a TV commercial flashed into tens of millions of American homes: a man sipped from a cup and then, behind a sour expression, proclaimed, "They can send a man to the moon! They can send rockets to Mars! Why can't they make a cup of coffee I like without caffeine?!"

The ad, which was successful, included a central theme very popular with a large segment of the general public: What, the Man in the Street asks, is space doing for *me?* The average citizen sees astronauts on the moon, knows they bring back boxes of lunar rocks, and has the vague idea that this is great for scientists, that it is advancing man's knowledge and helping accelerate the progressive march of technology. But all this seems remote, distant, and fuzzy—not directly applicable to the taxes he pays. He understands road taxes. Property Taxes, he knows, help finance his children's schools. But however small, the bite from his paycheck that funds space programs is almost incomprehensible.

Retired Congressman George P. Miller of California, former chairman of the House Committee on Science and Astronautics, addressed himself to this point this way: "America's space program can and does stand on its own feet, justified in its own right as an heroic manifestation of the evolutionary progress of humanity toward a higher and better life. It is justified by its success to date, and by its promise of continuing success . . . in the [expansion] of human knowledge of the universal environment in which our planet, Earth, travels. It is justified by what we are learning about our own satellite, the moon, which controls the ocean tides, and about the sun, which provides all of our energy and nourishes all life on earth.

"But not all our people believe this. Many good citizens and taxpayers . . . want . . . to know, 'What does the space program do for *me,* here on earth today, in tangible end-result returns from the investment we are making in it?'

"[These people] have a . . . right to ask what benefits are accruing to them *now* in return for their money. [So] it is important to tell this story."

The House Committee does this in part through the annual release of a report, "For the Benefit of All Mankind—The Practical Returns from Space Investment." Its introduction contains this statement:

"Many American citizens experience periodic thrill to the adventure of space exploration, yet wonder if their investment will pay a more direct return than the awe and fascination they have felt on the occasion of a rocket launch or a manned flight splashdown.

"The truth is that space research has produced an extremely broad range of concrete benefits, not only to the American citizenry, but to the people of many nations. The flow of 'hard benefits' has grown from a trickle to a stream, and it is widening to a river as expanding technology uncovers more and more ways of improving man's mode of existence on earth.

"The space age is still scientifically precocious—[just over] fifteen years old. Space research is not and should not be primarily oriented toward immediate direct benefit to the man in the street. The basic purpose of scientific research and advanced technology is long-term gain. But a striking aspect of space research is that a great many short-term dividends are being realized despite its strong orientation to the future. Advances in computers, miniaturization, electronics, exotic materials, and many other by-products have become part of our way of life almost without recognition."

As General Phillips puts it: "The influence of the space program has by now become so pervasive and is so quickly assimilated into our living patterns that we tend to take much of it for granted."

There are literally thousands of space spinoffs directly benefiting man's present life on earth. Physical examples are everywhere—in home and marketplace, in business and industry, in cities and on farms—and take many forms, such as new products, processes, techniques, and systems.

One immediately recognizable space-to-earth application area is consumer goods. On supermarket shelves everywhere there is a popular Pillsbury product, Space Food Sticks. Tasty, nourishing, and about the size of cigarettes, they provide a quick energy source for active people. They were developed as a snack that astronauts could eat with their helmets on. Similarly, shoppers can find various items—coffee, tea, soup, potatoes, even onions—in a freeze-dehydrated form initially designed to be carried in a spacecraft's cramped confines on manned missions in earth orbit and to the moon.

Those continually improving meals on airlines that span oceans and continents are largely the result of better cooling systems initially developed in the early 1960s for Gemini astronauts' spacesuits. The spacesuit coolers were modified into a refrigeration system for aircraft galleys that greatly enhances food preservation, yet requires neither batteries nor external power supplies.

Housewives can roast meats in half the time usually allowed because of a revolutionary pin inserted into the meat: it cooks the meat from the inside out. This spinoff specifically came from the technology necessary to meet high-heat-transfer requirements for space vehicles. The pin, using condensible vapor to transport heat, operates on the principle of the steam pipe, a simple device with several hundred times the heat-transfer capability of the best metal heat conductors. Meat can also be frozen faster with the cooking pin, since the pin can also work to "pump" heat from the roast.

At supermarket meat counters choice cuts of steak, chops, and other meats are wrapped in the same sort of transparent polyester film—a two-hundredth of an inch thick—used for NASA's giant balloon satellites, such as Echo. And many high-strength aluminum foils used to protect freeze-dried foods and

High energy snack—Space Food Sticks—was first developed as a special purpose food for astronauts on long space flights.

perishables are made of the same material used on early communications satellites.

The by-products of space are visible in nearly every room in a home. In the kitchen the hard temperature-resisting coating applied to certain cookware is fallout from the heat-shield technology developed to protect manned spacecraft reentering the earth's atmosphere through temperatures that reach up to 5000 degrees Fahrenheit. Similarly, spacecraft insulation so effective that when steaming hot coffee is poured into a tank covered by it, the coffee loses less than one degree of temperature *in a year,* has been adapted to thermos jugs.

Sealants prepared for caulking spacecraft seams now plug the gaps between shower tiles. Latex paints, invented because of the need in space for protection against ultraviolet radiation, are routinely applied to home walls.

In dens and family rooms the digital clock on the desktop and mantelpiece operate with a degree of accuracy not previously possible because they are being manufactured with a precision mechanism similar to one NASA designed to position and "fly" spacecraft models during wind-tunnel tests.

In the garage and around the yard high-energy-output batteries provide fast starts for portable power tools and sports equipment—again, thanks to technology derived from space flight requirements. These nickel-cadmium batteries, which can be recharged ninety to one hundred times faster than conventional ones, are also used as power sources in golf and baggage carts, portable medical equipment, photographic flash units, toys, and appliances.

Home builders are using more and more flat electrical conductor cables and low-voltage switching circuits produced by NASA more than ten years ago to decrease the size and increase the efficiency of spacecraft wiring. Because standard 110-volt wiring can be dangerous, it is normally encased in heavy metal cables and hidden inside the walls, where it is difficult to reach for repair or extension to any new electrical outlets and ceiling fixtures.

The new system brings household wiring to the surface. The wire is flat and thin enough to be concealed under carpeting, or even pasted on walls and covered with paint or wallpaper. It has been successfully installed in a new hotel at Marina del Rey, California, and builders have bought the equipment for experimental testing in a number of other states.

An adaptation of the Apollo fuel cell, which generated electricity through the chemical reaction of gaseous hydrogen and oxygen, is utilized now to supply electricity. A demonstration model of the new fuel cells is installed and operating in a condominium subdivision at Farmington, Connecticut. This technology is expected to have particular utility in small towns or areas remote from a central electric power system.

Using space systems management and engineering, as well as space-inspired manufacturing techniques and durable materials, a prominent electrical corporation in Pennsylvania has initiated a program of producing modular housing units that can be put together in a variety of shapes and sizes to suit the buyer. They are attractive and comfortable to live in, as well as economical to build and maintain. Many cities, subdividers, and builders have expressed interest in the idea.

Another company has adapted modular construction concepts to the production of cabin-type dwellings for migrant farmworkers in California's San Joaquin Valley.

There are space spinoffs in fashions, and fabrics that help make life more comfortable on earth—for example, specifications for astronaut undergarments led to the use of a specially designed texturized fabric, tailored according to a new underwear concept it provides unusual support without compromising flexibility, no matter how tight the fit. Designers are now manufacturing lines of these garments for consumer use.

Campers and sportsmen are also benefiting from the technological fallout. One fabric, originally designed as superinsulation, is made by laminating a special plastic material to a coating of aluminum only half a thousandth of an inch thick. It has found commercial marketing outlets in such consumer products as super lightweight blankets, sleeping bags, and outdoor apparel. The aluminized blanket is particularly popular because it can be used two ways: to keep the cold out by retaining body warmth or, conversely, to serve as an insulator to keep the heat out. It is waterproof, windproof, and surprisingly strong—but perhaps its most attractive feature is that it weighs only a few ounces, and although it can cover a king-sized bed, it can be folded up and stuffed into a coat pocket! It is also a bargain—some blankets being made from this material sell for under $4.

NASA's research into fabrics and materials has led to a superinsulating, vacuum-metalized nylon (taffeta) that is heat reflective, yet porous and machine washable. Consumer applications such as bedcovers, draperies, tents, and awnings are being developed.

America's firefighters also owe a debt to the United States space program for some of the new, safer apparel they are wearing. This includes thermal underwear, a coverall, chaps, trousers, jackets, caps worn under helmets, gloves, nonflammable boots, and a proximity suit which permits firemen to move closer to fires or even enter the flames if necessary. The nonflammable materials used in the garments were developed to insure maximum safety of moonbound Apollo crews in the spacecraft's oxygen-rich atmosphere. In fact, NASA has one of the most complete data banks ever compiled on the burning characteristics of different materials. In conducting its research, the space agency has maintained a close liaison and testing program with fire-fighting and fire-prevention associations to achieve practical designs.

Such other items as a new molded, nonflammable helmet for firemen—made from polymide resin and glass, employing technology first created by the Rockwell International Corporation for manned spacecraft use—is now in advanced stages of testing.

There are between two and three million preventable fires in the United States each year. They annually claim from ten to fifteen thousand lives, causing damage to property and other losses amounting to more than $10 billion. Because fire safety has been a prime NASA concern since the agency's founding in 1958, space research has led to many products and techniques, besides the new clothing, that should help lower grim fire-loss statistics in the 1970s and beyond.

NASA has developed several fire-retardant or nonflammable foams, paints, fabrics, and glass-fiber laminates. Paints that emit a flame-retaining gas when heat is applied have been studied by NASA's Ames Research Center and are being evaluated by the National Association of Home Builders. Foams developed at

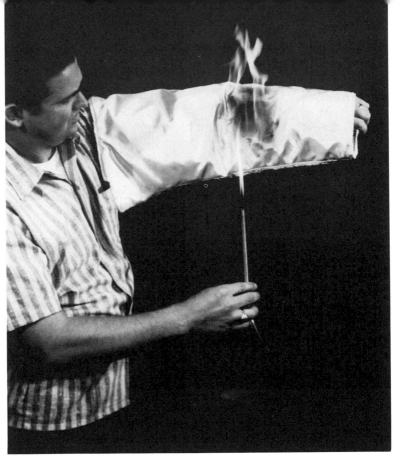

Fiery spinoff: New Beta fabric, developed initially for the space program, has fostered substantial improvements on insulated fireproof suits on earth.

Ames also retard fire propagation and may have excellent insulation properties. At the Johnson Spacecraft Center near Houston, extensive tests are under way to apply these and other materials to fire resistant carpeting, seats, head rests, paneling, curtains and fire walls.

Safety on the nation's roads—where many tens of thousands of Americans are killed and another 11 million injured each year—also is being improved through applications of space-related technology. For example an energy absorber, designed for astronaut couches on Apollo spacecraft, has been modified for use in highway guard rails. The Bureau of Public Roads reports that several states have installed this new system, which reduces a sixty-mph impact to the equivalent of a five-mph impact by dissipating heat and absorbing energy when an axial force is applied. The Ford Motor Company has an intensive program under way to incorporate this device into automobile bumpers that will be capable of withstanding a five-mph collision.

Freeway commuters in at least eighteen states have a much greater chance of reaching work and home safely today, thanks to a unique NASA-evolved project for preventing "hydroplaning" or skidding by aircraft on rain-drenched airport runways. Through research studies, it was learned that a grooving technique prevents formation of a continuous water film on runway surfaces, assisting pilots by insuring traction for their aircraft tires and permitting them to apply brakes safely. National Airport in Washington, D.C., was so impressed in 1967 and results were positive enough to expand the application to highways. State safety officers reported an 80-to-90-percent reduction in damage, injury, and death from skid accidents on roads that have been grooved.

Tires, too, are being made safer. An ultrasensitive fast-scanning infrared optical device that was first used by NASA for nondestructive testing of miniaturized electronic circuits is now being applied by the B. F. Goodrich Company for testing new designs in aircraft and automobile tires. The device produces a real-time cathode-ray-tube picture of the heat in tires as they spin rapidly on a special rig—up to 200 mph for car tires and as fast as 400 mph for aircraft tires. The camera is capable of reading the heat from 600,000 points on a tire every second, presenting an infrared "heat picture" of the tire, in which flaws or hot areas appear as bright spots.

When astronaut Gene Cernan's helmet visor "fogged up" during a space walk on a Gemini flight in 1966, NASA engineers came up with a special anti-fogging compound. It is now finding commercial application—on automobile windshields, and for motorcycle, diving, and firemen's helmet visors. It also has proved effective on plastic, aluminum, and other reflective surfaces where it is desired to maintain a fog-free state.

In another major contribution to automotive safety, a computer program designed by NASA to analyze the behavior of spacecraft structures under stress has been adapted for use in checking front suspension and steering linkages in a line of American cars and light trucks. These computer analysis techniques result in a 60-percent improvement in predicting the behavior of components under stress.

Waterway safety has been given a boost by a lightweight inflatable untippable, radar-reflecting liferaft. Used by the United States Coast Guard and marketed commercially as well, it was originally designed as part of the astronauts' onboard survival kits to use during splashdown upon their return from space. Made of lightweight fabric, the raft has an orange canopy that can be "spotted" by radar at distances up to fifteen miles—a major breakthrough for survival. The Coast Guard uses them aboard its icebreakers, and the Federal Aviation Administration is testing the raft for possible use on transoceanic aircraft.

Spinoffs from exploration of space have had profound and continuing effects on American education, and so have the communications satellites. An estimated 20,000–25,000 teachers are using spinoffs, for example:

- Instructors at a Sacramento, California, high school no longer must keep tedious attendance records. There is a ten-digit space-developed computer keyboard in each classroom, and the teacher merely punches the code number of an absent pupil. The signal is filed in a central computer, and at the end of the day a printed readout notes the attendance or absence of every pupil in the school.
- In California and New York City high schools teachers wear or carry a pen-sized alarm unit—a device that grew out of space miniaturization techniques—and if there is a disturbance, pushing a toggle switch on the pen sends an ultrasonic signal to the main office, where equipment identifies the signal's source and location so that help is quickly dispatched.
- In Brevard County, Florida—home of Cape Canaveral and NASA's Kennedy Space Center—the curriculum of a high-school credit course for adults includes basic earth, physical, and biological science information emanating from the space program.

Two million schoolchildren in every state get crisp up-to-date space briefings from a traveling NASA spacemobile that offers free lectures and demonstrations.

More difficult to measure is the overall impact of the Space Age, in the last decade and a half, on science-education practices. There was a great restimulation of interest in physics and new math subjects, for example, and processes of computerized and educational learning have been accelerated, and great masses of new knowledge, inspired by space research, is being infused into classrooms. On this point Dr. Lee DuBridge, former Science Advisor to the President, has said, ". . . The dawning of the Space Age has impelled Americans to seek to improve their schools. That alone may be worth the cost of all our space rockets."

Working with universities, NASA compiles the relevant Information produced by its programs into curriculum supplements made available to teachers. This project helps fill the gap between the appearance of new knowledge and its inclusion in textbooks, which often lag because of time required for preparation.

The general approach of NASA's elementary- and secondary-school programs is to offer teachers fresh information in useful formats. The teacher judges how and when to employ the new knowledge in the classroom. Central to this approach is a program of assistance, to institutions of higher learning, state and local school authorities, and professional associations in the conduct of courses, institutes, conferences, and workshops.

"NASA's greatest contribution to United States education, however," says Dr. Thomas O. Paine, former NASA administrator, "has undoubtedly been the information academic researchers have received from our direct involvement of the university community in the space program. What these scientists and graduate students learn in the pursuit of their research feeds back immediately into the teaching, publication and learning process, thus becoming available to the new student generation and technical community."

○
Chapter 19

miracles in medicine

In March 1972 a medical aide in the remote village of Allakaket, which lies above the Polar Circle in northern Alaska and has a population of only 125, needed emergency help for Sally Sam, a seriously ill eleven-year-old. Because many towns in Alaska have been equipped with ground station antennas that usually are able to bounce messages off ATS-1, NASA's first Applications Technology Satellite, the medical aide tried desperately to contact the United States Public Health Service Hospital in Tanana, a town one hundred miles south on the Yukon River.

For some reason the aide couldn't rouse Tanana, but he did get through to an ATS control station in Mojave, California, which in turn relayed the message by satellite 22,000 miles above earth to Anchorage and from there to Tanana—all in minutes.

The Public Health Service physicians there diagnosed the illness as acute appendicitis, and fifteen minutes later an evacuation aircraft, with doctor and patient aboard, took off from Tanana for Allakaket. The little girl was taken to the hospital for surgery and recovered fully.

A few months earlier, in one of the finest, most modern medical facilities in the world—California's Stanford University Hospital in Palo Alto—twenty-five-year-old Mrs. Mary Phillips was bleeding to death, and a team of doctors had run out of conventional ways to save her life. Over a five-week period they had given her forty-six pints of whole blood and sixty-four units of plasma and had performed nine operations, but they could not halt the persistent abdominal bleeding.

Dr. H. Ward Trueblood, chief resident in surgery, called NASA's Ames Research Center in nearby Mountain View, explained the situation, and asked for help. A team of specialists studied the unusual problem and offered an unprecedented solution: Mrs. Phillips should be put into a pressure suit of the sort test pilots wear to prevent them from blacking out during high-speed aerial maneuvers, for it applies pressure that counters the draining of blood from the brain and torso.

The suit stopped Mrs. Phillips' internal hemorrhaging overnight by reducing the difference in pressure between the blood in her arteries and the tissue outside of them. Her blood thus could coagulate naturally, and she was soon able to return home to resume a normal life.

Mrs. Phillips and young Sally Sam are alive today because of technological spinoff benefits from the United States space program. The advanced processes used in saving them are only two of thousands of research discoveries and engineering innovations—developed as astronauts headed for the moon or as spacecraft speed to distant planets—being dramatically adapted to at-home medical applications. To help speed this transition, NASA has assigned task forces of specially trained personnel to work directly with medical researchers and channel complex data into everyday use.

"The weightless world of space is one of man's few remaining frontiers," says Dr. Charles A. Berry, former NASA director for Life Sciences. "It is the science of aerospace medicine that makes it possible for man to explore this new frontier. . . . Much of the derived understanding can be applied to ground-based physiology and medicine."

For a heart attack victim, for example, every second between the initial strike of pain and professional help can mean the difference between life or death. More than 60 percent of the deaths occur within an hour after the first attack, yet until recently the ambulance transit period was time lost in terms of diagnosis and treatment.

The Space Age is changing that: in an ingenious system designed to check astronauts' heart action, electrodes are applied to a patient's body so an electrocardiogram (EKG) can be flashed from the ambulance via a radiotelephone link to the hospital while the patient is en route. Reading the EKG at a hospital console, the doctor's advance knowledge of the patient's condition enables him to make necessary preparations before the patient arrives, thus saving precious minutes. The system has been tested successfully in Los Angeles and other metropolitan areas.

In this era of ever-rising medical-care costs, the increasing use of space-related electronic devices and techniques promises substantial reductions in hospital expenses by freeing skilled personnel from routine patient-watching duties. For example, a tiny sensor and radio transmitter, developed under NASA sponsorship, has been modified automatically to monitor infants and comatose adults suffering from windpipe obstructions. When injury or disease causes blockage of the upper-respiratory passage, surgeons frequently must perform a tracheotomy—the insertion of a small tube in the throat to bypass the nose and permit free breathing. If the tube becomes clogged, breathing stops and brain damage or death will result within two to four minutes.

A full-time nurse must therefore constantly watch the tube and take any necessary corrective action immediately. By noting subtle differences in the temperature of air passing through the tube, this new sensor activates an audible or visible alarm within ten seconds of change. Such a signal can be connected to a console at a nurse's station, allowing one attendant to keep watch over dozens of patients.

A similar system is in use on an experimental basis at the

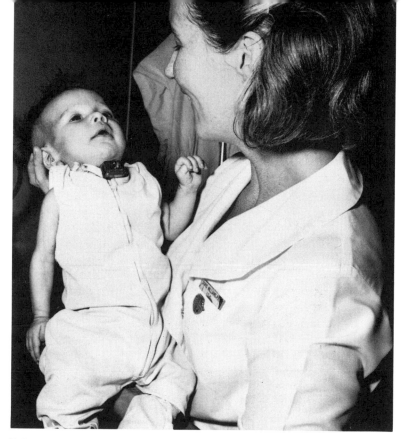

Life-saver: Tiny sensor and radio transmitter, placed at the throat of an infant, sounds an alarm if the child has any difficulty in breathing. The device was originally developed in space research.

Children's Hospital Medical Center in Oakland, California. Delicate sensors are attached with microminiature connectors to newborn infants with respiratory ailments, and the transmitter sends a pulsating signal to a central nurse station. If the baby has trouble breathing, the signal is interrupted, instantaneously warning nearby attendants. To the experienced ear, the quality of the continuous tone transmitted can provide valuable information concerning the infant's respiration.

In the intensive care unit of a Miami hospital small battery-powered electronic devices manufactured by aerospace companies are strapped to patients' arms and legs. Through them such vital physiological information as temperature and blood pressure can be transmitted from as many as sixty-four patients to a single nurse at a monitoring console.

"This is the type of benefit which is of particular assistance when we note that many of the Veterans' Administration hospitals do not have the necessary personnel to care adequately for the patients," says Florida Congressman Louis Frey.

Sight switch, triggered by the movement of a patient's eyes, flashes signal to a nurse at a monitoring station instantly. Techniques such as this were initially designed to aid busy astronauts in space.

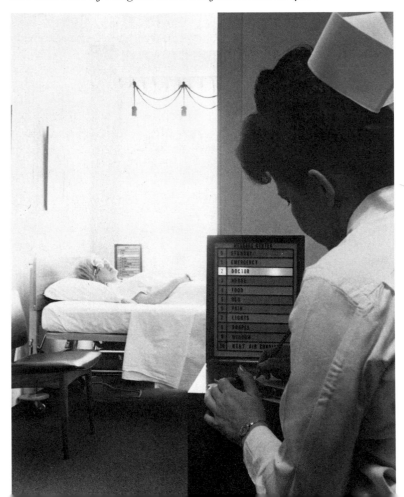

At a major New York hospital the commercial model of a NASA biotelemetry unit is in use in a cardiac monitoring station. Oscilloscopes display the heartbeat patterns of patients wearing wireless heart sensors and transmitters, and an alarm-signal device is activated wherever a patient shows the beginnings of abnormal activity. A polygraph records and stores accumulated heart records for each patient.

The prime advantage of a wireless biotelemetry system is that a person can perform normal functions, even mild exercise, while his heart signals are constantly monitored at a central facility, freeing trained staff members from having to stay at patients' bedsides constantly.

Around the country strange-looking contraptions and paraphernalia—originally designed exclusively for moon bound astronauts—have been adapted to a number of practical, if unusual applications for the sick, injured, lame, and paralyzed.

At the Texas Institute for Rehabilitation and Research, for instance, handicapped children learn to walk in a sling-support outfit invented at NASA's Langley Research Center to help acquaint spacemen with the one-sixth gravity conditions of the lunar surface. The rig consists of a walkway, tilted off horizontal, on which patients exercise while suspended sideways by belts and pulleys. Various modifications of this sling are used to help bedridden patients who find it difficult to retrain unused muscles to walk again.

"Such devices take the load off the heart, the respiratory system, the lower limbs, and relieve the strain on all of the dynamic 'subsystems' of the body," says Dr. James Gaume of Long Beach, California.

Doctors at the University Medical Center in Kansas City have found a unique use for astronauts' helmets: respirometers for children. Youngsters always found the conventional rubber mouthpieces used for the collection of exhaled breath uncomfortable and difficult to keep in place. At times this affected the accuracy of the data on oxygen consumption.

Helmets made to order to solve the problem are equipped with an air inlet and outlet, a rubber seal around the neck, and a suction pump to permit constant circulation of fresh air, collect the exhaled breath, and draw the combined fresh air and exhaled breath into an oxygen analyzer. And the kids, of course, love to use them. They have turned a troublesome chore into a pleasant experience.

Surgeons also use spacesuit helmets instead of surgical masks, which in addition to comfort, are more sanitary and therefore cut down even more on the possibility of surgical sepsis. Some doctors wear treated surgical garments that bacteria cannot penetrate. And a number of hospitals are adapting the ultraclean laminar airflow techniques, implemented by NASA for assembling spacecraft, in the use of superfine filters that purge the most minute dust particles from the air during surgery.

One of the most extraordinary space instruments to be applied to medical purposes is a "sight switch" initially developed to aid busy astronauts on long-distance flights. It has been successfully converted for quadriplegics—persons with use of neither arms nor legs. With this instrument, a patient can easily manipulate a motor-driven wheelchair by the movement of his eyes. The novel switch operates on the principle of infrared reflection from the eyeball. An infrared light source bounces light off the white of the eye into a photoelectric cell which carries the message to a control activator. When the eyes are moved sideward, one eye reflects the light while the pupil of the other eye absorbs it. The resulting imbalance of voltage controls the direction of the wheelchair. Such a device also can be used, for example, to turn pages of a book, flip a thermostat on or off, and control radio and television sets.

This specialized wheelchair has been tested and favorably evaluated by the Rehabilitation Institute of the New York University Medical Center, more recently it was tested and refined for use by paralytics at the Rancho Los Amigos Hospital in Downey, California—ironically, the same city where the Apollo spacecraft that carry astronauts to the moon and elsewhere were built.

Dr. Howard Rusk, head of the Rehabilitation Institute, estimates that 100,000 persons in America who have use of neither their arms nor legs will be able to use this wheelchair.

Even Apollo's specially designed windshields have been put to work on health care problems. They are used in the treatment of severe burns. These three-by-four-foot curved "shields" are

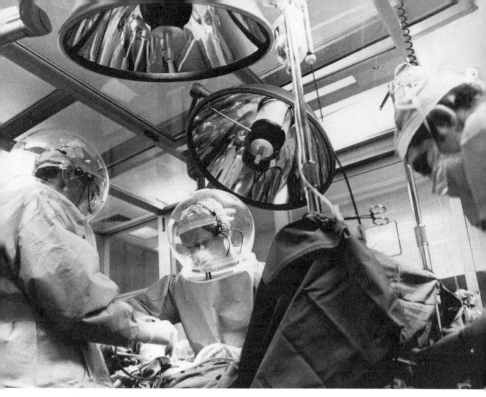

Super-clean surgery: Doctors at St. Luke's Hospital in Denver perform a hip-joint replacement in a new, clean, room facility which helps lessen the danger of infection to the patient. Surgeons wear astronaut-type helmets and garments that are impermeable to bacteria.

suspended over patients, allowing freer movement while a constant temperature is maintained under their surface. Water loss through evaporation, one of the major factors in the healing process, also is cut down through use of the constant-temperature shield, and doctors have found in many cases where the shields are applied, pain-suppressant narcotic drugs are not necessary.

"Of particular importance in burn cases is the comfort of the patient, and the need to keep him warm," explains Dr. Richard Grossman, specialist at the Sherman Oaks Hospital Burn Treatment Center in California. "Since the shield does not encompass the patient, it is much easier to change dressings and for the attending physician to determine the progress of the recovery.

More than one thousand patients in a dozen centers have been cared for using the new shield, and doctors report it to be a valuable aid," Grossman says.

Similarly, a dry immersion bed designed for research on astronauts' metabolic rates in the weightless conditions of space provides relief for victims of severe skin burns or ulcers. Patients using the bed lie on a waterproof sheet and remain dry while "floating" in water heated to body temperature. Because the patient is buoyantly supported, the pressure is evenly distributed, rather than exerting maximum pressure on any single part of the body. The device is used for people in long-term bed confinement cases to prevent and to treat pressure-produced skin ulcers and to treat massive skin burns.

The technology fallout affects maternity wards as well. The company that developed the "glass sandwich" to prevent windshield-fogging has transformed its expertise into the production of cradle-warmers. It monitors the condition of premature infants and maintains the proper temperature inside a laminated plastic bassinet. Several hundred such units have been purchased by hospitals and clinics.

At many of the nation's leading medical and hospital centers additional space-spawned ideas are in various stages of research, experimentation, and laboratory testing. NASA specialists are working with Stanford University School of Medicine personnel, for example, on a dramatic new use for sonar that, it is hoped, will expose secrets about the functioning of the human heart.

Sonar machines, which emit and receive high-frequency sound waves, can measure precisely the amount of blood pumped out by each heart muscle contraction, something standard monitoring devices don't do. This technique can be used to screen patients with known or suspected heart disease, and it can be used to follow the progress of the healing process in patients recovering from a heart attack or open-heart surgery.

Equally exciting is the research centered on NASA-developed blood-pressure sensors so small—some with diameters of less than five one-hundredths of an inch—they can pass through a dog's artery into the heart with no bad effects. At Harvard Medical School such devices have successfully tested blood pressure with unprecedented accuracy. With an ordinary hypodermic needle the sensor is inserted into an artery; attached to a thin flexible tube, it is maneuvered into the left ventricle to measure inside the artery and heart without disturbing the blood flow. Permanent implantation of a sensor with a transmitter in a human body would allow continuous monitoring of a patient as he moved about freely. And its small size would make it particularly useful in treating babies.

At the St. Louis School of Nursing and Health Services a revolutionary "muscle accelerometer" has successfully undergone extensive testing to measure minute muscular tremors in the human body. Patterned after a "momentum transducer," it was produced by NASA to measure micrometeorite hits on spacecraft and is capable of recording impacts as faint as that made by a grain of sand dropped one inch. This may help doctors in early diagnosis of conditions such as Parkinson's disease.

Among the most intriguing potential applications in medicine of space-spawned devices is a new electrostatic camera that produces moving or still "instant pictures" with no processing. It can focus on a patient in critical condition and keep vital photographic records instantly available for physicians. Transducer-transmitters that relay intestinal data are currently in use, and doctors now anticipate a battery-powered television system small enough to be swallowed, which would transmit pictures *from a patient's stomach!*

Scientists and doctors at The Johns Hopkins University's Applied Physics Laboratory have demonstrated a new heart pacemaker that is longlasting and much smaller than conventional units, and which uses electrical and electronic components initially designed by NASA for use in spacecraft. Previous pacemakers had several drawbacks, including battery failure and comparatively large size and weight, which caused patient discomfort.

Because of its small size (approximately a third the volume, half the weight, and half the thickness of other pacemakers), the new rechargeable equipment is virtually hidden within the patient's body. This pacemaker incorporates a modified version of space satellite power cells, specially designed for implant in the body. The basic concept for this cell has been proved in more than ten years of space use.

Aids to the blind and deaf are also being derived from space research. The principle of alternating panoramic fixation used in satellite camera and lens systems has been applied to the development of new glasses with multidirectional lenses. General Data Corporation, which makes instruments for spacecraft, is doing research on an electronic sight aid for the blind. The company also has adapted the small electronic sensing devices used in spacecraft for another use—the restoration of hearing to the deaf by surgical implantation. Eye surgery performed by a pinpoint of intense light from a laser beam is fully accomplished, and the Kollsman Instrument Corporation, for one, indicates that the laser beams can also be used in tumor removal, retinal welding, and brain surgery.

There is new hope for narcotics addicts because of a sleep analyzer used by NASA to check the sleep of astronaut crews on Skylab missions. It combines the electroencephalograph and electrooculograph signals from the sleeping patient's brain and eyes to record the depth and hence the quality of sleep. In addition to its possible use as a diagnostic tool in the treatment of addicts, it is seeing earthbound use in other forms, namely as a detector of fatigue danger for pilots, tower operators, or others with critical jobs.

The space-medicine spinoffs often promise less pain and discomfort for the patient. Ultrafast drills, with minute ball bearings developed through research for satellite equipment, are available to dentists for nearly painless dental work.

The ear oximeter eliminates the need for repeated blood-pressure checks or needle-in-the-vein blood tests of people suffering from shock. Designed at NASA's Ames Research Center, the oximeter, attached to a patient's ear, gives doctors blood-pressure data when the conventional methods of attaining it are not feasible. It measures the blood's oxygen content by noting red- and infrared-light absorption in blood circulating through the ear. The onset of shock is accompanies by a reduction in both the quantity and the oxygen content of blood flowing through the earlobe. The resulting changes in the infrared absorption and blood pressure detected by the oximeter cause the device to set off an alarm. Doctors can then take prompt corrective action. The aural oximeter is proving specially important to patients with leukemia, to provide an early detection of physiological shock.

The list of usable medical spinoffs from space—tangible ones that are helping save lives and mend bodies now, and others that are being converted from research laboratories to practical uses almost daily—includes scores of products, instruments, techniques and systems . . . lightweight foam splints for broken arms and legs, . . . new alloys for improved artificial limbs. . . . electronic devices that convert audio sounds to vibrations that can be "heard" through the sensitive fingertips of the deaf and blind . . . digital computer processes that clarify medical X-rays. . . .

Yet for all the benefits so far tapped from space technology to date, they have only scratched the surface of what realistically can be expected in the near future. The nature of space flight has been largely experimental, the emphasis having settled on engineering, on solving the overwhelming technical problems of merely getting to and from the moon easily. But now the thrust is shifting more to the exploitation of space for the betterment of man.

The three manned earth orbital Skylab missions typified this new direction: top priority was given to medical experiments. NASA sought to find out how prolonged time spent in space—up to two months—would affect the astronauts. The metabolic and cardiovascular systems were measured in depth, and serious investigation of the orbital effect on nutrition, hematology, immunology, neurophysiology and pulmonary functions also was undertaken.

Primarily, scientists are interested in identifying the precise mechanisms that change the human body's chemistry when earth's gravity is absent. This knowledge is important not only in planning for future long-range space flights, perhaps to other planets, but it can also contribute to a better understanding of life processes which are basic to treating human illnesses here on earth.

○
Chapter 20

a new order of technology

The extraordinary levels of reliability and performance demanded for space program hardware and systems, the highest ever set by man, have created a new order of technology—one that is of tremendous significance to business and industry. To meet these demands, research and development projects call upon an unprecedented, combined national resource that includes NASA, other government agencies, private industry, and university-based knowledge and expertise within virtually every field in the science and engineering disciplines. This massive effort, so broad in scope yet so minutely exact, provides breakthroughs and state-of-the-art advancements that not only serve as a source of new technology for the space program, but also constitute a vast new resource for the nation's industrial capability.

To help channel these data to practical uses on earth, in the early 1960s NASA established an Office of Industry Affairs and Technology Utilization, and it serves as a clearing house and actively promotes the widest possible application of space technology to the civilian community.

The Space Act of 1958, which extablished NASA, stipulated that knowledge generated in national space programs must be made available to the American people. The law specifically required NASA to disseminate information as widely as possible.

This is done in a number of ways: the publication and distribution of *Tech Briefs* and support packages, compilations on technology related to general subjects, surveys in special fields, and periodic conferences.

Every six months the Space Agency issues a nine-hundred-page industrial "best-seller" called "Patent Abstracts Bibliography," which tells of nearly two thousand NASA-owned inventions available for licensing. Scientific and technical aerospace reports and abstract journals are written, edited, and published.

At six regional dissemination centers across the nation, computers store NASA's vast technical files—including the results of more than $45-billion worth of contract work performed for the Space Agency by tens of thousands of companies and other organizations. All these data are available to businesses, regardless of size, for a nominal fee.

"Our product is somebody else's technology," says Dr. Daniel U. Wilde, director of the New England Research Application Center in Storrs, Connecticut. "Every information search we perform for some company helps NASA fulfill its charter and may lead to a profit for the company."

When approached by a business concern with a technical problem, center staff members, all experts, examine the situation in detail and suggest areas to search for applicable information. In addition to NASA's comprehensive data bank, the centers store data gathered by several technical societies. "All of these resources are available in our searches," says Dr. Wilde. "We can go through millions of pages of technical reports in a matter of hours. It's our job to see that NASA-developed technology is transferred to the public. The technology was bought or developed with public money, so it belongs to the public.

"We're not charging for technology. We're charging for the service we provide. The fees are small compared to what companies would have to spend if they tried to develop the same information on their own. The expense would be needless if NASA has already done the research." The network of centers is operated under the Technology Utilization Office.

Nearly twenty thousand individual technology items have been identified as having specific application potential for commercial users. Of these, over three thousand have been published in the NASA *Tech Briefs*, of which more than 40 million copies have been distributed.

TECHNOLOGICAL WINDFALL FROM SPACE

A small Idaho company that makes oscilloscopes spent $190 for a computer search of NASA's technical data banks, and gleaned the fruits of five years of oscilloscope research done under government contract by a major aerospace company. The data included hitherto overlooked techniques for building a special instrument. Put to use by the company, this information helped add more than $100,000 in sales over the next few years, doubling its business.

In addition to the hundreds of space-to-earth uses of specific products and devices—in the home, in medicine, in education and safety, etc.—the great technological windfall from space has many broader applications.

Municipal government—like American industry, medicine, and agriculture—has borrowed technology and methodology from space experience, and taxpayers are the direct beneficiaries. "Mobilizing modern science, technology, and management to accomplish bold ventures in space is clearly far simpler than better organizing the extraordinarily complex human interactions that comprise a modern metropolis," says Dr. Thomas Paine.

"Few cities today have the managerial structure and resources to take early advantage of technical opportunities, much less to foresee new possibilities and deliberately bring about needed technical advances applicable to urban systems. This is anachronistic . . . and can only lead to deteriorating services and soaring budgets.

". . . New federal and local management institutions must be created, based on the realities of today's metropolitan areas. Major resources must be administered under close control, orchestrating the best talents of universities, industry, and government to apply the great power of modern science, technology, and management."

Dr. Paine adds that NASA's spectacular advances have proved to be a source of encouragement so that the nation can tackle its complex human problems with greater confidence on a bolder scale. "If America can go to the moon," he says, "it can indeed do much better here on Spaceship Earth."

Specific applications to the cities and their systems and departments include improved detection and locating equipment for ferrous and nonferrous underground piping; new technology in solid-waste management; improved pavement patching matériel which uses certain kinds of waste; drug-detector equipment, and a psychomotor response tester for use in traffic departments and educational institutions.

Consider a single application of a NASA-developed procedure, this one for analysis of paint chips, to the science of urban crime detection. Criminal labs often have to identify the automobile in a hit-and-run accident from only a bit of paint scraped at the scene of the crime. Generally this is done by visually comparing the paint chip with sets of standard paints supplied by car manufacturers. The Space Age technique uses a spectrophotometer and obtains more precise results. It is being used successfully by the Sheriff's Office in San Mateo, California, and in other communities.

A Sacramento aerojet firm performed a systems analysis on the overall problems of crime and delinquency under a contract with the State of California. One outcome of the intensive study was a crime-detection technique called "neutron activation analysis," which compares clues found at the scene of a crime with samples taken from a suspect. The "nuclear fingerprints" could be used, for example, in comparing tire marks, hair samples, and marijuana. Matching probabilities of the neutron-activation analysis were said to be 99.999 percent.

Computer-control systems have been adapted to municipal administration in a number of metropolitan areas. Los Angeles turned to space program management techniques for help in meeting the increasing demands on police, fire, and ambulance service. The City of Angels is implementing a command-and-control system to provide rapid pinpointing of field forces, computer dispatching, automated status displays, computerized information files, individual communications for hazardous-duty personnel, and automatic transmission and signaling for emergency vehicles.

Among new metropolitan transportation systems San Francisco's much-publicized Bay Area Rapid Transit (BART) took ideas and methods from aerospace research to help build what has been called the nation's most modern municipal rail transportation project. BART links more than thirty stations in three counties via a seventy-five-mile system that includes the world's longest underwater transit tunnel—beneath San Francisco Bay.

While the technical achievements derived from the space program contribute in many ways to better life on earth, there is exciting potential too in the mastery and management of the multiple skills that have produced such achievements. Large aerospace programs, such as the Apollo moon landing missions, have revolutionized business and industry management techniques as well as those of many government agencies. They motivate and unify many highly intelligent and energetic people of diverse technical skills; they keep track of a myriad of parallel processes, identify problems quickly, and continually adapt to changes needed as experience produces more knowledge. These new management techniques may well have application in solving the socioeconomic problems confronting the world.

As an example of such an application, Kansas City recently completed a $250-million international airport, using a NASA management system. This vast project's construction was directed by a fully equipped management information center installed on the eighth floor of the city hall, patterned after and closely resembling the facility at the Kennedy Space Center that used to manage the Apollo program.

NEW MANAGEMENT TECHNIQUES

The techniques for directing the massive projects undertaken by thousands of minds in a close-knit, synergistic combination of government, universities, and industries represent one of the most important of all spinoff benefits. These techniques are potentially the most powerful management tool in man's history, changing the way civil servants, scientists, and managers approach virtually any task.

The American computer industry does about $10 billion worth of business annually. American computer exports have increased 1400 percent in the past decade, and almost every major computer system is made in the United States; the industry employs more than 800,000 people. The space program has been instrumental in helping accelerate the boom-growth of the Computer Age.

The assimilation of scientific data for each stage of a space flight . . . the design and production of virtually every component of a spacecraft . . . the precision control of that craft in flight . . . and the storage, classification, and retrieval of data from every manned or unmanned space mission have produced tremendous advances in computer technology.

For example, the rigid weight and volume requirements of spacecraft have pushed ahead, by years, the development of microelectronics. Thousands of circuits are routinely compressed into chips smaller and thinner than a human fingernail. The miniaturized components are finding their way into scores of commercial outlets, electronic calculators only a little larger than a pack of cigarettes, for one.

Special computer "software" developed for the wide range of operations in the Apollo program have been adapted for use with computers in air traffic control, industrial process control, engineering design, automation of hospital services, and sophisticated medical diagnosis.

Computer data-processing techniques and programs developed by space projects have enabled airlines to provide immediate flight information and reservations systems, insurance companies to improve their accounting and investment services, and other businesses to handle transactions involving more than 20 million items daily.

Another example: more than seventy industrial firms, universities, laboratories, and government agencies in the United States are using NASA Structural Analysis (NASTRAN)—a computer management system—to solve their structural engineering problems. NASTRAN has been adapted to nearly two hundred applications, ranging from suspension units and steering linkages on automobiles, to the design of powerplants and skyscrapers. Scores of other uses are in planning stages. NASTRAN is a general-purpose digital-computer program originally conceived to analyze the behavior of elastic structures in the space program, since one of its major uses was in the design of the space shuttle.

The payoff for industry is that spacecraft, missiles, and moon missions aren't the only large problems around. An automobile body is pretty complicated from a stress analysis standpoint. So is a machine tool subjected to vibration. And many industrial stress and vibration problems are thought to be too complicated for any type of detailed analytical treatments. NASTRAN now provides long-awaited answers where none previously existed.

The applications of computerized numerical control and digital logic to the machine tool received their greatest impetus from the space program's metal-machining requirements. To send men to the moon and satellites far from earth, a system that assured the automatic control of machinery by means of programmed instruction stored on punched cards or tape, replacing the operator jigs and manual controls, was essential to meet "zero defects" standards of the highest priority.

"The marriage of numerical control, the digital computer, and machine tools is one of the stunning technological innovations of our time, ranking with nuclear power and space flight itself as a third great development of our generation," says Willard F. Rockwell, Jr., the American industrialist and Chairman of the Board of Rockwell International.

Not only will more than three-fourths of all machined parts be produced by numerical control, with tremendous improvements in productivity, which means lower costs as well as greater reliability, but the computer is also enabling management to make sounder decisions on the basis of faster, more accurate, more complete, and experienced information. Specific uses of numerically controlled tools can be made in cutting metal, material handling, assembly, welding, fabrication, inspection, quality control, computer graphics, and machine drafting and plotting.

SPACE AGE MATERIALS

From the Stone Age through the Iron and Bronze Ages and into the second half of the twentieth century, material technology has advanced to meet industry's demands and spread its benefits to world welfare and economy. The Space Age has speeded this progressive march to a prodigious pace.

New metallic materials developed for space vehicle applications include high-strength stainless steels, strong weldable aluminum compositions, and corrosion-resistant weldable titanium alloys. Precipitation-hardened stainless steels are used in the manufacture of military and civilian aircraft, submarines, and ships. The United States Army has produced an all-welded lightweight aluminum-alloy amphibious vehicle. And titanium alloys are widely employed in chemical and petrochemical installations, especially in oil refineries, where corrosive chemicals quickly destroy ordinary steel valves.

One of the critical design criteria for space vehicles is that structural materials must have the least weight, the greatest strength and stiffness, ability to operate in a wide range of temperatures, and submit to fabrication processes such as forming, machining, and welding. This has resulted in different uses of materials and the extraction from them of superior performances that would have been unimaginable a few years ago. Such materials include beryllium, titanium, stainless steels, super alloys, refractory metals, aluminum, magnesium, and high-strength steels. All are applicable to innumerable nonspace uses in almost every metalworking industry for all sectors of the economy—transportation, mining, farming, refining, chemicals, electrical, communication, construction, and road building.

From Apollo have come vacuum- and radiation-resistant lubricants, used to reduce the friction of moving parts, that offer higher lubricity and can be used in high-speed turbines and other rotating equipment. Technology spinoff is leading to the emergence of high-temperature lubricants for jet engines and industrial machinery such as rotating kilns, driers, and roasters. In nonmetallic materials, research conducted on resins, polymers, and plastics to develop materials that would be immune to ionizing radiation in space are adaptable to a wide variety of industrial uses. High-strength polyurethane and polyamide adhesives offer suitable bonding agents for many space and nonspace uses. A variety of aircraft structures being fabricated today are taking advantage of these new adhesives.

The requirement for cryogenic insulation of rocket tanks—to contain fuels at temperatures to −300 degrees Fahrenheit—has resulted in a highly effective, lightweight spray-on foam insulation. The material can be used for insulating cold-cargo containers, tankers, prefabricated housing construction, orthopedic splints, and furniture packaging among other things.

NEW PROCESSES IN MATERIAL HANDLING

Material-handling processing has also improved vastly too in the Space Age. For example, American industry can now weld aluminum alloys that did not even exist a few years ago; it is welding complex shapes, materials a few thousandths of an inch thick, and in areas considered inaccessible to humans. All of these techniques were pioneered by the Apollo program. An automatic "skate" welder, used on Saturn moon rocket stages, can travel the length of the meridian tank weld, making a consistently perfect weldment which can be used in large structural projects, like shipbuilding. Tungsten inert-gas welding, necessary for the strength and nonpermeability requirements of space, is available for structures exposed to high pressures, extreme temperatures, and corrosive effects—anything from small pressurized containers to submarines.

POWER GENERATION

Also directly attributable to the space program are the major improvements in power-generation technology that promise far-reaching effects for remote-area power generation, the automotive industry, deep-sea technology, communications, home and commercial power sources and air pollution.

Developed under NASA sponsorship to assure life support in the spacecraft's sealed environment, there is a wide variety of applications of fuel cells for improving man's living conditions. Fuel cells are producing electricity aboard one-man submarines, and powering experimental spot welders, golf carts, tractors, and fork-lift trucks.

The Allis Chalmers Company has demonstrated a farm tractor and a passenger car electrically powered by fuel cells using hydrocarbons and air. Cars run by fuel cells will ultimately make a very significant contribution to an economic elimination of air pollution.

Hoffman Electronics Corporation manufactures a solar-powered radio that derives from Vanguard satellite solar cells. A commercial wristwatch is powered by a mercury battery developed to operate a timing mechanism for an Explorer satellite. Batteries made for spacecraft use have often been modified for medical use. One can be sewn into a patient's body to overcome cardiac defects, still others are used in an emergency call system on the Los Angeles freeway, and for a telephone system in South Africa. Silver-zinc batteries, developed for military rockets, are useful in deep-submergence vehicles and submarines.

United Aircraft Corporation, which designed the Apollo's fuel-cell power plants, is putting this technology to work to produce a marketable gas-powered fuel cell that promises to be important for home and commercial building supply.

Even rocket engines are being employed as new power sources. The Commonwealth Edison Company of Chicago—one of the nation's largest electric companies—has teamed with Rockwell International to build the first "space-power" machine, a project that has long interested utility companies worldwide; a rocket engine, which instead of producing thrust for boosting a vehicle into space turns water into steam to drive a turbine-generator. And it is being done by a process that virtually eliminates emissions of particulates and other pollutants—and without the noise normally associated with large rocket engines. Installed at Commonwealth Edison's facility at Joliet, Illinois, this new unit has been designed to generate eleven thousand kilowatts of electricity—enough power to serve about ten thousand homes.

But these space-to-earth technology benefits are merely representative examples, one small chapter in a large and growing book. Thousands of others are in various stages of transfer, from planning to full implementation.

"As we succeed in transferring aerospace technology to other nontechnology-intensive industries, or meet pressing public sector problems, we help ourselves," says Jeffrey T. Hamilton, Director of NASA's Technology Utilization Office. "We have better products, cleaner air, better health, and in the process we can create technology-intensive higher-skilled jobs in industries that don't have them."

The examples cited here are not vague possibilities for a distant tomorrow, nor are they blue-sky propaganda. Often unplanned, unpredicted, they are nonetheless real and are based on existing or impending technological capability.

"They are in fact *extra* dividends, which are a fallout of ingenious application of space flight experience by business, industry, commerce, science, government, the medical profession, and the academic community," cites a Congressional Report on space spinoffs.

"Those dividends already paid, coupled with those in sight for the near-term future, affect practically every facet of human convenience and concern. They promise continuing and increasing return on our space investment for the benefit of mankind on earth today."

So to those who say, "We can go to the moon, why can't we . . . ?"

The answer is, "We can!"

Chapter 21

pro and con

. . . It's a fundamental law of nature that either you must grow, or you must die. Whether that be an idea, whether that be a man, whether that be a flower or a country. I thank God that our country has chosen to grow.

—ASTRONAUT GENE CERNAN

"The committee judged the promises and offers of this mission to be impossible, vain, and worthy of rejection: that [it] was not proper to favor an affair that rested on such weak foundations and which appeared uncertain and impossible to any education [*sic*] person, however little learning he might have."

This "thumbs-down" excerpt is from a report written in 1491 by the Talavera Commission, which had been considering a proposal by a brash young Italian adventurer named Christopher Columbus. Had Queen Isabella not overruled the commission, the history of the Western world would have been drastically altered.

There are direct parallels between the Talavera Commission and the Columbus expedition nearly a half-millennium past, and modern critics and the current national space program. "Some people ask, 'Why should we spend this money to explore space when there is so much to be done here on earth?'" says Congressman Olin E. Teague. "Well, there was plenty to be done in Europe when Columbus left it. And there is still plenty to be done there. If Columbus had waited until Europe had no more internal problems, he would still be waiting, but the opening up of the New World did more to revive the European culture and economy than any internal actions could possibly have done."

Inventors and their inventions, ideas that were revolutionary and far ahead of their time, rapid advances in the progress of technology—all too often these have been ignored or laughed at, criticized and lobbied against. Astrophysicist and space visionary Arthur C. Clarke, commenting on a committee set up in the Middle Ages to decide whether it was worth developing Mr. Gutenberg's press, once wrote, "After lengthy deliberations, this committee decided not to allocate further funds. The printing press, it was agreed, was a clever idea, but it could have no large-scale application. There would never be any big demand for books—for the simple reason that only a microscopic fraction of the population could read."

Whenever Leonardo da Vinci strayed from the artist's brush to the designer's drawing board, his brilliantly prophetic ideas and concepts were scoffed at, not by the uneducated alone, but by some of the best minds of his times. Galileo was ridiculed and imprisoned for daring to suggest the blasphemous notion of heliocentricity. Threatened with excommunication from the Church, he recanted to the Inquisition, but was said to have muttered as he left the room, a free man, "It's not my fault that the universe isn't geocentric. I couldn't help it."

When Congress was asked to appropriate funds for the exploration and eventual settlement of the West, Daniel Webster voted against the idea, saying that it would be a waste of the taxpayers' money because, as everybody knew, that barren territory had nothing but scrub cactus, deserts, high mountains, and savages. In 1867, when the Secretary of State had the uncommon foresight to purchase Alaska for $7 million, he was called everything from "Spendthrift" to "Idiot," and the acquisition came to be called "Seward's Folly."

Not until his later years, after the wonderful practicality and usefulness of his inventions was proven, did Thomas Edison gain respect and admiration. The Wright Brothers' early attempts to build an airplane were laughed at—everyone knew it was ridiculous to think that something heavier than air could fly.

"Some [space] critics point out that it has not been demonstrated in detailed certainty that [the space program] will be economically rewarding," says Senator Howard W. Cannon of Nevada, a member of the Committee on Aeronautical and Space Sciences. "History is full of examples of major ventures that would have been delayed for generations or perhaps never undertaken if such total proof-in-advance had been required. . . ."

In the late nineteenth century someone asked Sir William Preece, chief engineer of England's Post Office, if he cared to comment on the latest American invention, the telephone. He is said to have replied, "No, sir. The Americans have need of the telephone, but we do not. We have plenty of messenger boys."

When a constituent asked Senator Cannon what good space was doing, he cited the following: "Many years ago the British scientist Michael Faraday was lecturing on one of his many discoveries, when a member of the audience somewhat irreverently interrupted him, asking, 'Yes, but what good is it?' Faraday's reply was, 'What good is an infant?' "

Robert Anderson, president of Rockwell International Corporation, adds, "From the very beginning, engineers and scientists have been involved in building this nation. As much as any group they are the spirit of America. Science and engineering tied this nation together with the railroads, the automobile, and the telephone. It was the airplane, product of science and engineering, that figuratively eradicated any sectional difference in this nation. Radio gave it one voice, television gave it one image. This is part of technology's contribution to the spirit of America.

"Surely the world is changing; the scientists and engineers helped change it for the better. They've earned their place in the American society. They are contributors, both materially and to the spirit that has made this nation number one. Consequently, we're no longer going to listen in silence to sandal-shod nonsense about returning to the bucolic life of nineteenth-century simplicity. America is an urbanized society, and our job is not to turn the country back to rural simplicity, but to raise it to new high standards of living for everyone."

The nation's space program is in its infancy. Despite the tremendous benefits of its far-ranging missions already being enjoyed on earth, despite the unlimited potential it offers for a better future for all mankind, there are many vocal, powerful, and erudite critics who believe space should be abandoned.

Congressman H. R. Gross of Iowa, for example, has said, "This business [space] ought to be phased out and halted until we can get on our financial feet in this country."

British historian Arnold J. Toynbee observed, "In a sense, going to the moon is like building the pyramids or Louis XIV's palace at Versailles. It's rather scandalous, when human beings are going short of necessities, to do this. If we're clever enough to reach the moon, don't we feel foolish in our mismanagement of human affairs?"

Congressman E. G. Shuster of Pennsylvania has said, "We are told that we need to spend $3 billion . . . for our space program, but in the same breath we are told that there is not enough money . . . to finance a public water and sewer program. In my own congressional district, 65 percent of the homes do not have public water and sewer systems. There is something wrong when we are more willing to put a man in outer space than to put him in a modern bathroom."

Congressman William Lehman of Florida has said, "While people have thrilled to our accomplishments in outer space, no one wishes to see space program increases paid for by cutting back vital health and educational programs. We have been told that important domestic programs are to be cut back because there is no money.

"These [space] programs represent a sadly misguided system of priorities which this Congress must change. Space programs are not more important than health programs. Funds for space exploration must not be increased while funds for education are decreased. We must reject the President's space program increases until we have . . . alleviated the hardships of life on this planet."

Such charges may sound appealing to constituents caught in

the dollar squeeze. When prices are rising and taxes are being raised, it has become accepted practice, at least since Apollo 11, to direct angry charges at the space program and suggest that if we cut off all funds and directed them to more down-to-earth needs, all domestic problems would miraculously be solved.

A close examination of these charges, however, reveals their flaws. Perhaps the most popular critical theme has been, "The space program is taking money that could be better spent for more pressing programs, social welfare, for instance." To those unaware of the benefits of space, this line is enormously appealing. Statistics, however, prove its fallacy.

In 1969, when man first landed on the moon, the United States government spent $65.2 billion on "social action" programs, including income security [welfare], health, veterans' benefits and services, education and manpower, community development and housing. Only $4.2 billion were allocated to space. In 1970 the ratio was greater: $75.4 billion were funded for "social action" programs, $3.7 billion for space.

"In 1971," says Congressman Teague, on . . . "health, commerce, transportation, education and welfare benefits, and other services and agricultural development, we spent approximately $95 billion. In 1971 we spent about $3 billion in space research and development."

In 1972 more than $60 billion were allocated to welfare programs, $16 billion to health programs, more than $10 billion for veterans' benefits and services, nearly $9 billion for education and manpower, and about $4.5 billion for community development and housing. The national space program budget was just over $3 billion. In other words, $100 billion of the federal budget were targeted for social and domestic action programs while $3 billion went to space. (In terms of the Gross National Product, space expenditures account for about .3 percent.)

As Congressman James D. McDevitt of Colorado explains, ". . . 46 percent of our national budget already is allotted to domestic programs related to our human and physical resources, while only 1.4 percent is spent on space. What can we accomplish with 1.4 percent that we could not do with 46 percent?"

On this specific point Senator Frank Moss of Utah, says, "If you took X number of dollars from one source, there is no assurance that they would be applied to Y program. Now I believe in funding all human programs on earth. We shouldn't slight this.

"But with a trillion-dollar economy . . . we ought to be able to afford less than one percent of it to scientific and technological advances. In so doing, we are not only helping provide answers for some of the problems today, but we are insuring a more secure future for the next generation, and the next."

And Senator Charles MacMathias of Maryland adds, "Arguments that we should not provide funds [for space] because it does not feed hungry children, provide needed housing, etc., are specious. To use this argument as the basis for opposition . . . to space . . . misses an essential point. For it appears to be based on the false logic that implies that we can never pursue a second- or third-priority goal as long as our top priority is not completely achieved. And that is hardly a prescription for a balance of priorities."

A second line of attack is, "Look at all the money we are wasting on space!" But not a cent is spent on space—it is all spent on earth. A spacecraft can orbit or fly to the moon, but it is designed, developed, and manufactured on earth. More than 92 percent of all space funds have been spent in the United States, in all fifty states. During the late 1960s, the peak of space years, the economic impact was felt by more than 20,000 American companies at the prime and subcontractor level—and they employed nearly 300,000 people annually.

Furthermore, a substantial portion of the dollars spent on the space program returns to the city, state, and federal governments as corporate and personal taxes. In the past ten years the taxes returned to these governments amounted to over $3 billion, and they contribute to the social and domestic programs.

"Solutions for our domestic problems," Congressman Teague comments, "require a large amount of money along with new ideas and technology. This in turn is dependent upon a strong and expanded economic base. To promote an expanded economic base requires adequate support of research and development to create the new and more efficient processes, materials, tools, and techniques needed to solve our day-to-day problems.

"Those nations . . . which do not persevere in research-and-development programs are . . . those which [also] fail to develop an economy which adequately feeds, clothes, educates, and houses its peoples."

"You can make sort of an analogy [between the space situation and] the energy situation today," Senator Moss has said. "Everyone knows we're running short of natural resources. The Administration says we must increase research-and-development funding to find additional energy resources. If we don't do this, our standard of living will rapidly [decline].

"You can apply that philosophy to space too . . . if we don't build upon the solid foundation we have created to date, we will dry up all possibilities of reaping future dividends from our program.

"How does one evaluate the benefits of the space program? It is not easy. How does one evaluate the benefits and return on investment of basic research? For example, who can assess the value that basic research had in discovering vaccines that have virtually wiped out polio? . . ."

Consider the irony that the very social and domestic problems confronting Americans today might, in the long run, best be solved through the application of space-related technologies and systems. "To meet the pressing social problems of our times," NASA's Dr. Fletcher says, "requires above all a sound economy operating at a high level of employment to generate the tax revenues required at all levels of government. To maintain such an economy in a competitive world, we must increase our productivity year after year, decade after decade. The only way . . . to keep increasing our productivity is through advancing our technology."

Rocket engineer and space prophet Dr. Krafft Ehricke says, "It is often claimed today—in fact, it is a very popular theme— that the space program is not socially relevant. It is not socially responsive. The average American fails to see how space technologies can help improve socioeconomic conditions on earth, or how it can help us preserve our precious environment, etc.

"In reality, however, the space engineer is the social planner's strong ally, the environmentalist's greatest ally."

The multifaceted impact of the technology dollar, Ehricke says, produces and will continue to produce the breakthroughs that make possible control of environmental pollution, advances in medical science, extension of education to illiterate areas everywhere, building of cheaper but better housing; above all it is these breakthroughs that generate the dynamic industrial technologies, and thus new products, markets, and jobs.

The goal is not to replace social action programs with technology programs, but neither is it to abandon technology because of the need for social action programs. The greatest need is for a combination, so balanced that finding solutions to earthbound problems can be accelerated by technological advance.

"Space *is* socially relevant," says Dr. Ehricke, "but only if we recognize it and use it."

"We should not be too harsh with critics," says Dr. Fletcher, "for they do render an important service by reminding us repeatedly of the need to gain public understanding and support for the space program. For in our zeal to forge ahead in technical and scientific fields, it is possible to lose sight of the debt we owe to the great American public that so steadfastly sustained us during the early years.

"We should not—we cannot—blame the public for being more interested in taxation, social security benefits and farm subsidies. These financial outlays affect us directly and almost daily. The average American—like you and me—has his priorities too. But he needs to be informed if he is to arrive at an intelligent decision about how to rank his priorities.

"The burden of explaining—and even convincing—the public of the necessity for continuing a well balanced space program for the 1970s and beyond lies with those of us so directly involved," says NASA's Dr. Fletcher. "In this effort we should be comforted by the knowledge that, despite the dissenters of yesterday, and there have been many, the character and final judgment of Americans is clearly recorded in the progressive and pioneering history of this country."

Part Six

the
future

$\bigcirc$
Chapter 22

the new continents

The date has not yet been set—it depends on a number of factors, of which the most important are the future placement of the space program in the order of national priorities, and man's continued progress in developing his ability to live comfortably and work effectively in the space environment. The precise time is still in question, but it is inevitable that Americans sooner or later will operate large stations in earth orbit for relatively long periods, and then larger space bases over longer time spans.

Initially the space shuttle will serve as such a station—in the early 1980s—but even with the technological capabilities proved through the Skylab program, not until man can stay in space routinely will he be able to fulfill his destiny. With Apollo and other early projects we have explored. With the shuttle we will have begun to exploit the infinite wealth and potential of extra-terrestrial resources. With stations and bases we can colonize space and use it to gain the greatest direct benefits for earth.

It is reasonable to assume that the United States will put up its first station during the 1980s. Although the specific design and the final choice of missions are still being studied, we know a lot about the first station: how it will look, the systems it will require, its operation.

NASA says it will be a semipermanent facility in earth orbit with an operational life of at least ten years. It will probably be modular, so additions can be made to the basic structure even after it has been placed in orbit and serviced by the shuttle. The core station could weigh up to 100,000 pounds and have five decks, two of which would be for living, eating, sleeping, and administering the station; one would receive and house supplies ferried by the shuttle, another would contain subsystems—such as environmental control, to maintain a pure, breathable atmosphere, and water management, to reclaim waste water for reuse—and the fifth area would be a laboratory for a wide variety of experimental activities. This five-level first module can have other modules attached to it and here we would have specialized experiments or developments.

The station as envisioned will provide comfortable living accommodations for a twelve-man crew, which will include specialists to maintain and operate the vehicle, and nonastronauts, scientists and engineers, to make observations and run the experiments. Crew members can be rotated on a regularly scheduled basis, but probably would stay in space six months or even longer.

To keep morale high and to stimulate creative and effective work, considerable care will be taken in decorating and stocking the living facilities. Food preparation and service will be as close as possible to that on earth; fresh-frozen meats and vegetables will be available. Quiet private quarters will be provided in which the men or women can work, write, or read. Adequate personal-hygiene facilities will be provided so the crew can be clean and well groomed. A central wardroom could serve for dining as well as conferences on coordinating station activities.

The crew will be able to live and work in the same "shirtsleeve" environment as on earth. The living quarters will have the same constituents and pressure as air at sea level—this artificial gravity, in the weightlessness of space, is created by swinging the station like a pendulum counterbalanced by a spent rocket-booster stage, which will be connected by long cables. There will be areas as well of zero gravity in which scientists can conduct experiments in the pure vacuum of space. Large solar cell arrays or a nuclear power system will provide electric power.

The first station will thus be the space equivalent of an explorer's terrestrial base, like those in the Antarctic. It is being designed so that only a small percentage of the crew's time will be required for housework. Most of the time will be available for experimental work.

The mission potential in a space station is really limited only by man's imagination. On earth these can include ecological studies in depth, pollution research, communications, transportation, weather forecasting, education, resource survey, forestry research, geology, materials processing, and technology spinoff.

Scientific experiments will be in oceanography, meteorology, biology, medicine, astronomy, physics, and chemistry, and operational projects could include interplanetary launchings from space. The station could also serve as a supply base, a rescue craft base, and a maintenance installation for unmanned satellites. Later, work on the station will be directed to the use of solar energy on earth, and for determining means to develop worldwide illumination techniques from orbit.

Many of these efforts will complement and advance today's work on unmanned satellites and on Skylab, and additional projects to be carried out when the space shuttle is operational. In other areas the station will offer a unique platform for programs that can be accomplished only by a manned crew in a large laboratory-like space environment for relatively long periods.

One such area is materials processing, or manufacturing. There have been preliminary experiments in this direction on Skylab and they will be continued on the shuttle, but man will not truly be able to assess his ability to manufacture anything in the weightlessness of space until the station is a reality.

But it has been proven that certain manufacturing processes can be done better in the absence of gravitational fields or the presence of a vacuum. Materials that will not mix on earth—oil and water, certain metals—will mix in space, perhaps making possible the production of new and improved materials, the more precise manufacture of products, the new material processing. Such work could be done in a module attached to the space station that would house a workroom with processing chambers, tools, equipment, and storage space for raw materials and gases. A large airlock could provide access to the space environment or be opened to space to produce a large sheltered vacuum chamber.

What, specifically, could be done in so unique a laboratory? Development of metal foam, for one. Space offers the promise of producing stable foams from a wide variety of liquefied materials and gases; hence it should be possible to manufacture a foamed steel with the weight of balsa, but with many properties of solid steel. This process is impossible on earth because the weight of the liquid metal causes the gas foam bubbles to rise to the surface before cooling can take place. In weightlessness, gases will remain entrapped and produce a spongelike material.

Similar techniques can be used to mix materials of such vastly different densities and properties as steel and glass. Composite and foamed materials should yield lighter and stronger material for basic study, probable industrial applications, and even future spacecraft construction.

Single crystals grown on earth are limited in size only by outside forces or contaminants. In an immaculate zero-gravity space environment there are no limits to potential growth. Oversized crystals, grown of the right material and with the proper impurities controlled, could be used as very large power transistors, and if they are pure quartz as optical blanks for near-perfect lenses.

Another aspect of space-materials processing that intrigues scientists and engineers is levitation melting. Suspending a specimen (i.e., levitation) is important because it offers the possibility of melting materials without contamination from so much as a crucible or any mechanical restraint. Metallic materials and structures can be shaped by the manipulation of surrounding electromagnetic fields with resulting perfect shapes. The virtual absence of gravitational forces makes this a natural process for investigation in space. This process isn't practically possible on earth.

Gravity conditions require more current to levitate the material than is needed to melt it, and it cannot be cooled unless it rests on something. In space, however, metallic material is easily levitated and held in place, and the current can be increased for softening or melting. Manipulation of the magnetic fields in a coil system permits the specimen to be moved about without its touching anything. It can be melted and resolidified, therefore, without becoming contaminated or deformed.

Such a method would permit production of perfect spheres for use as ball bearings that would inaugurate a new order of mobility for wheels. Uniform alloys could be produced because maximum intermixing of constituents is possible, as would be refinement of metals to high levels of purity.

Metal foam, crystal growth, levitation are only three examples of the uses of space for revolutionizing certain materials and manufacturing processes, and many other applications are being studied. One expert suggests that the absence of gravity will permit production of drugs, cheaper and purer, and preparation of such processes as electrophoretic purification of vaccines, and incubation processes for biologicals.

Glass of a quality far superior to that produced on earth can be produced as containerless solidifications, and high-quality lenses manufactured for lasers and optical instruments. Physical and chemical fluid processes are being researched for the possibility of processing in space, as are such metallurgical processes as metal-matrix composites and eutectic and monotectic alloys of controlled structures.

"It's like the old West," says Dr. Krafft Ehricke, drawing an analogy. "The government had to finance the railroads to cross the country before Americans could migrate west. Today the government must finance space transportation systems to make possible the manufacturing of superior items beyond earth's influence. Private industry already is interested, but it cannot go to work in space until the government has built the transportation system."

Private industry participation could, in fact, result in the eventual operation and ownership of space manufacturing facilities by corporate bodies under an organizational setup similar to that under which Comsat (the Communications Satellite Corporation) was established. To foster this idea, NASA would make available to private industry some of its present and future resources—the shuttle, space workshops, and station facilities—at a mutually agreed-upon, reasonable price. An era of orbiting factories producing superior products impossible to produce on earth will routinely be manufactured at great cost savings.

A semipermanent space station will greatly enhance the direct economic returns from exploiting the potential of orbiting observation platforms. The enormous benefits derived from operations weather satellites, for example, are well known. But consider the forecasting potential of professional meteorologists using advanced instrumentation and making direct observations of weather phenomena from a space station! Imagine the contributions to earth resources a geologist could make, or an agricultural expert, a forester, a hydrologist, an oceanographer—they could interpret and analyze raw information obtained in flight, selecting what to transmit to earth and thus reducing the burden on communications systems; recognition of patterns, which can be difficult to include in a computer program, could be done by all station personnel.

The station would be a fixed space research facility for conducting scientific investigations to extend man's knowledge of the nature of the universe. A large telescope module, similar to the Skylab telescope mount and operating in conjunction with the station, could be designed to observe ultraviolet and infrared in addition to visible radiation from the stars and planets. Modules equipped with different instruments could observe solar phenomena or study X-rays from celestial objects for extended periods. A telescope compartment and instrument area on the module could be pressurized to accommodate an astronomer. The astronomer could conduct observations, perform experiments and handle necessary maintenance and repairs.

A space station offers life scientists the opportunity to study life processes under conditions impossible to duplicate on earth. Significant advances would result from long-term study in space of plants and animals, including man, by doctors, biologists, botanists, and others. Man's presence is desirable for most such experiments because biological organisms and their life processes are inherently variable and complex, requiring immediate interpretation, the extraction of information from incomplete data, and the rearrangement of experiments. Equipment able to manipulate soft tissue and handle lively animals as well defies automation.

In a biological laboratory designed for experiments with plants and animals and evaluation of human physiological and medical condition, special compartments will house study specimens and have separate environmental control systems. Experiments will permit a small staff of experts to perform comprehensive physical analyses of life in space which undoubtedly will have wide application on earth for the betterment of life.

One laboratory, a deck on the station or a separate module attached to it, could house high-energy physics and cosmic-ray studies. The heart of this facility might be a superconducting magnet surrounded by chambers for tracking incoming charged particles, such as protons and electrons, that originate outside the solar system and travel at very high speed. Physicists believe they hold answers to some of the basic questions about the nature of matter.

With an orbiting station as the centerpiece, various structures can be assembled in its immediate vicinity—for example, huge antennae, too large to be launched fully extended, could be orbited folded or in pieces, and crew members could contribute to chores in space and assemble, deploy, calibrate, service, and repair them. Because of an antenna's size and the precision of manipulation required, fully automated deployment of large antennae carries a high failure risk as the failure of solar-cell arrays on the first manned Skylab mission attests.

Orbiting structures can be used for any number of purposes—one of the more obvious of which would be to improve communications systems, while one of the bizarre, envisioned by Dr. Ehricke, would be the deployment in geosynchronous orbit of a complex of mirrors, nearly a square mile in diameter. With this, says Dr. Ehricke, "You will have a means of illuminating 36,000 square miles of [earth] surface without disturbing ecological effects. This can be done by solar reflection, involving no power consumption and therefore no pollution on earth.

"The area would be illuminated to about ten times the full moon's brightness on a clear night, and several times the full

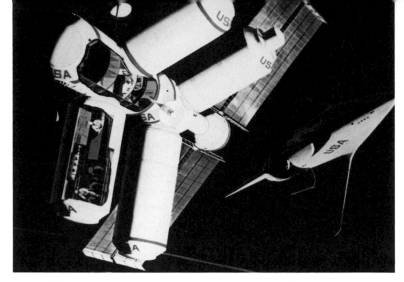

Orbital home: An artist's version of how a modular space station of tomorrow, serviced by a space shuttle vehicle, right, may look.

moon's brightness when in the presence of a cloud cover. Can you imagine the effects of such a phenomenon? Benefits range from greater public safety in urban areas to agricultural advantages in being able to work in the cool of night in tropical regions."

The space station and the follow-on space base will serve a major function as a transfer point in orbit. Planners have long been fascinated by the possibility of using such a facility as a departure site for missions being launched to higher orbits, to the moon and into deeper space. The station would serve as fuel and supply storage and transfer facility as well as an assembly point for vehicles too large to be launched directly from the earth's surface.

For example, supplies and crewmen could shuttle to the station, where they would transfer to another vehicle perhaps nuclear-propelled, for transit to lunar or geosynchronous orbit. In this mode massive payloads could be transported to the moon. Similar flights to the planets could begin in orbit, after assembling and stocking the rocket-spacecraft combination.

"Provided it has been proven economically feasible and desirable, the space station could eventually be developed into a large space base in low earth orbit," says Dale Myers, former associate NASA administrator for Manned Space Flight. "The base would be constructed by clustering space stations and other specialized modules. It would provide a large space laboratory of common equipment and modules where non-astronaut scientific personnel would be able to conduct a variety of scientific experiments.

"Initially, the base would accommodate approximately 50 persons, including the operations team to perform command, control, service, and maintenance functions. Growth to a 100-man capacity would be possible. To incorporate the extent of flexibility toward future growth that may be required for a national research and operations center, the space base would be modular in construction, so that it could be remodeled or expanded in orbit, using the logistic capabilities of the space shuttle."

Myers says the spectrum of activities to be conducted at the station and base cannot yet be predicted. "This situation is analogous to the one that exists whenever plans are being formulated for a major new earth-based laboratory. In fact, history shows that unexpected uses, discoveries and payoffs often outweigh the planned payoffs on the frontiers of science and technology."

Another view of a possible modular space station, built from Skylab-type workshop tanks, and serviced by a space shuttle vehicle, lower left, is depicted in this artist's conception.

○
Chapter 23

the infinite promise

What lies beyond the space shuttle, beyond space stations, beyond lunar colonies? What will follow once the solar system's planets have been visited, first by robot spacecraft and later by man? What benefits can be reaped from the endless seas of space surrounding earth—twenty years from now, thirty, fifty?

Will illiteracy, ignorance, and hostilities be eliminated? Through the magic of satellites will man have created a united and monolingual world? Will pollution be a forgotten unpleasant phenomenon of the mid-twentieth century? Will we have harnessed the sun's boundless energies for daily use on our home planet?

One view of the future has been offered by Dr. Thomas O. Paine, former NASA administrator: "I am certain that by the end of this century men and women will be living and working in extraterrestrial space stations and in small colonies on the moon. The first few will have reached Mars. By then the first children will have been born outside the earth's biosphere and terrestrial life will be adapting itself to new worlds. A new sociology of extraterrestrial societies will evolve to suit growing communities of scientists, engineers and spacemen striving to create new human habitats in novel environments beyond earth."

Others envision vast complexes permanently stationed above earth and including huge manufacturing centers, medical centers and hospitals, even hotels with dynariums in which twenty-first-century tourists could cavort in waterless gravity-free three-dimensional "swimming pools."

Who would dare challenge such bold predictions?

When the United States launched a tiny metal ball into space on January 31, 1958, who could have predicted with any accuracy what would happen in so short a time? Man on the moon? Absurd! It will never happen in *my* lifetime! Remember?

Will we find other life?

American anthropologist Margaret Mead says, "Once you raise the question that other land than this earth is possible to live on, that other places are possible places to found colonies, or that there may be other living creatures somewhere, you have changed the whole place of man in the universe. You've altered everything. This involves a considerable reduction of human arrogance and a tremendous magnification of human possibilities."

"There are those who believe earth is a lonely and isolated island surrounded by barren, hostile wastelands of space," Dr. Ehricke comments. "They miss the point. Earth is not isolated in space, any more than a ship is isolated at sea. A ship is isolated when it is stranded on land. On the sea it is in its element. Such elements belong together—a ship and the sea, and earth and space.

"Space is not isolating us. Space is opening the gates. The message should be that we see earth, not as our world, but as a part of a greater system of worlds that has now become accessible to us.

"We must not revive medieval earth-is-the-center-of-the-universe syndromes, cosmic neoisolationism. The real message is, earth is unique in some respects, not in others. Factories, power-generating stations, minerals—these are not unique to earth. We can perform these functions from space. We can use space as an industrial sink. Why do these things on earth at risk of destroying the truly unique features—the bioatmosphere, the good soil, the oceans, and most of all the mantle of life? We cannot reconstitute life. We cannot create life on earth. We can only destroy it. But we can also strive to protect it by limiting the extent of terrestrial industrial elevation," says Dr. Ehricke.

"Earth is the only luxury passenger liner in a convoy of freighters loaded with resources. These resources are for us to use, after earth has hatched us to the point where we have the intelligence and the means to gain partial independence from our planet—and where the time has come to convert our earth from an all-supplying womb into a home for the long future of the human race, finally born into the greater environment of many worlds.

"Take the moon. Here, we have a raw material source, a huge motherlode, if you will, orbiting in the sky. We could move industry to the moon and space-truck clean materials back to earth. It doesn't matter if mankind pollutes the moon a little, or exploits its resources. The business of earth is life. The 52 million square miles of land on earth are the most precious real estate in the solar system. Let these other dead worlds be a place for industry," says Dr. Ehricke.

"Among our sister worlds that are reasonably accessible to us—the moon, Mars, Mercury, the asteroids—are treasure houses of minerals and all the other elements that human minds and tools will need to supply mankind on this precious, living planet—and to make earth truly the Acropolis, the upper city of our solar system."

In addition to its potential as an industrial sink and as a supplier of raw minerals and other resources, space also has the potential to be a power source or as a superpower relay station to help alleviate energy shortages on earth.

Concepts in this area already being studied include a system in which electrical power plants could be built wherever there is an energy source. Stations in remote areas could house nuclear plants so that ecological damage would be minimal. Instead of stringing costly power lines, the power would be beamed to orbiting satellites by microwave transmission facilities. The spacecraft would relay it to a microwave receiver and a nearby power-conversion and -distribution plant that would handle the power without chemical waste and with low heat discharge. A metropolitan area such as Los Angeles could thus use power generated hundreds of miles away, perhaps in the Nevada desert, theoretically without the environmental disadvantages associated with large power plant systems.

Another, perhaps nearer-term application of space technology to the solution of earth energy problems could be through a satellite's discovery of subsurface geothermal power sources, such as the heat generation that causes geysers and volcanic flows. Scientists believe that such sources, which have long mystified man, could be pinpointed and better understood when "viewed" by sensitive instrumentation aboard such unmanned satellite systems as ERTS or from such manned laboratories in space as the space shuttle. If sensors can lead to the detection of minerals underground, it is reasonable to assume that they can also provide clues about geothermal power pockets all over the world.

Ultimately, however, in view of the energy crisis, and if man continues his present energy consumption rate for an appreciable time, "importing energy from the sun could become a future necessity," says Richard B. Marsten, director of communications programs for NASA. A great deal of energy could be provided, he explains, by "utilizing the solar power falling on earth, but the systems to do this are limited by clouds, night, real estate requirements, and transportation needs. Extraterrestrial systems which intercept sunlight have virtually unlimited power potential."

The Solar Energy Panel, formed by the National Science Foundation and NASA, has reported that if sufficient money were spent on research, by 2020 solar energy could be supplying up to 35 percent of the total building heating and cooling load, 30 percent of American gas fuel requirements, 10 percent of liquid fuel needs, and 20 percent of the electrical power demand. The panel believed that if the research were successful, solar power for heating buildings could begin to be available commercially within five years, for cooling buildings in six to ten years, synthesizing fuel from organic materials in five to eight years, and generating solar electric power in ten to fifteen years.

Direct building heating and cooling, which is provided from oil and gas, accounts for about 25 percent of all American energy consumption. According to the Solar Energy Panel, if large-scale solar energy use were realized, the United States could be saving $3 to $4 billion annually on oil and gas by 2000.

NASA has already issued study contracts for assessing the feasibility of harnessing solar power and channeling it through space for use on earth. The key to such a program, says Marsten, is the development of microwave technology for transmitting energy between extraterrestrial solar collection stations in orbit and on earth. Many power experts consider it only a matter of time until the technology can be developed, and to maintain life quality on earth, such development is inevitable, for the sun is the sole source of energy that is completely safe, pollution-free, and inexhaustible.

Besides the international space missions projected for the space shuttle, some forward-thinking leaders believe there may be other commercial applications for this revolutionary transportation system. One potential use was suggested in a speech to the Society of Experimental Test Pilots by Najeeb Halaby, former chairman and chief executive officer of Pan American World Airways.

"The 'Impossible Dream' we insist on dreaming," he said, "is that of the space shuttle as an air transport, carrying passengers from New York to Tokyo in 45 minutes, or from Los Angeles to Rome in 40.

"The key to practical application of the space shuttle is basically the same as the key to developing commerical aircraft: reduction of the cost per pound of payload. Since it takes much less energy to put an object in earth orbit than to fly the same object across the United States, there is no reason why the cost of space shuttle operations should not become as low as or lower than that of jets.

"I can visualize a rocket-powered vehicle with 100 passengers—not counting stewardesses—launched vertically in suborbital trajectory at the precise azimuth for its intended destination, reentering the atmosphere without excess g-forces, gliding unpowered to an altitude where its conventional jet engines will be started, then landing on a runway like a conventional airliner."

It is obvious that the potential uses of space for the betterment of life on earth are virtually limitless. Such applications will be contained only by man's imagination and his resourcefulness in exploiting them.

"Earthbound history has ended," American philosopher Earl Hubbard says. "Universal history has begun. It is this nation's destiny to help find many worlds for man and to bring to all men on earth a new hope—a new future in a universe of infinite diversity and opportunity."

"Beginning with this century," Dr. Ehricke observes, "civilization started outgrowing terrestrial means and resources. The trend of this process is irreversible. The extent will grow over the decades and centuries ahead. Nothing short of a total collapse of the driving and catalyzing motivations in human evolution can change this. Such collapse is more likely to occur if we refuse to expand. I cannot imagine a more foreboding, apocalyptic vision than the fate of a mankind endowed with cosmic powers and condemned to solitary confinement on one small planet. Without developing a planet-oriented—indeed, a cosmic—attitude, humanity in the long run does not have a future worth looking forward to.

"The first three-dimensional civilization will soon be born—a civilization in which earth becomes the center of a vast and growing sphere of human activity encompassing surrounding space and other worlds.

"In space," Dr. Ehricke continues, "we can build needed structures that dwarf anything man has ever built before. In space the architectural visions of future generations will last for the life of our star.

"Space orbits, not only other celestial bodies, are the new lands of our time and in the decades and centuries to come. Space stations are the historical but comparatively modest first step on a journey that knows no end.

"The earth from here is a grand oasis in the big vastness of space."
—Astronaut James Lovell, in orbit around the moon.

"The twenty-first century will be the age of the great solar system clippers. They will change the complexion of human civilization as profoundly as did their globe-spanning forerunners in centuries past.

"Perhaps most importantly, space is the only major challenge of our time that is not borne out of past acts of ignorance, indifference or man's inhumanity to man. It is not a call to rectify past mistakes. Space opens new horizons beyond earth and offers new beginnings in the ways we can manage this precious planet, with all the attendant aspirations, hopes, opportunities for creative action, for bringing the human family closer together and for contributing to a better future for all.

"Herein lies the ultimate greatness of space flight. Here is a planet to manage from space," Dr. Ehricke goes on. "Beyond lies the titanic wilderness of our planetary system to be explored and cultivated. By expanding through the universe, man fulfills his destiny as an element of life, endowed with the power of reason and the wisdom of the moral law within himself. And space is only the tip of the iceberg. Beneath lies a vast base of industrial, technological, scientific and spiritual excellence.

"From immediate, specific benefits to the people of this nation and the world—to the promise of a humanity which calls this solar system its home base—the challenging is as unlimited and enduring as space itself."

Index